Exercises in Basic Ring Theory

T0224076

Kluwer Texts in the Mathematical Sciences

VOLUME 20

A Graduate-Level Book Series

The titles published in this series are listed at the end of this volume.

Exercises in
Basic Ring Theory

by

Grigore Călugăreanu

"Babeş-Bolyai" University,
Cluj-Napoca, Romania

and

Peter Hamburg

Fernuniversität GH,
Hagen, Germany

KLUWER ACADEMIC PUBLISHERS
DORDRECHT / BOSTON / LONDON

A C.I.P. Catalogue record for this book is available from the Library of Congress.

ISBN 978-90-481-4985-8

Published by Kluwer Academic Publishers,
P.O. Box 17, 3300 AA Dordrecht, The Netherlands.

Sold and distributed in the U.S.A. and Canada
by Kluwer Academic Publishers,
101 Philip Drive, Norwell, MA 02061, U.S.A.

In all other countries, sold and distributed
by Kluwer Academic Publishers,
P.O. Box 322, 3300 AH Dordrecht, The Netherlands.

Printed on acid-free paper

All Rights Reserved
© 1998 Kluwer Academic Publishers
Softcover reprint of the hardcover 1st edition 1998
No part of the material protected by this copyright notice may be reproduced or
utilized in any form or by any means, electronic or mechanical,
including photocopying, recording or by any information storage and
retrieval system, without written permission from the copyright owner.

This volume is dedicated to the memory of my father and our professor

George Călugăreanu 1902 - 1976

to my beloved family

Mara, Ilinca, Manole and Daniela

and to the memory of my mother

Zoe Călugăreanu 1905 - 1996

Contents

CONTENTS

Preface

Each undergraduate course of algebra begins with basic notions and results concerning groups, rings, modules and linear algebra. That is, it begins with *simple* notions and *simple* results.

Our intention was to provide a collection of exercises which cover *only* the easy part of ring theory, what we have named the "Basics of Ring Theory". This seems to be the part each student or beginner in ring theory (or even algebra) should know - but surely trying to solve as many of these exercises as possible independently.

As difficult (or impossible) as this may seem, we have made every effort to avoid modules, lattices and field extensions in this collection and to remain in the ring area as much as possible.

A brief look at the bibliography obviously shows that we don't claim much originality (one could name this the folklore of ring theory) for the statements of the exercises we have chosen (but this was a difficult task: indeed, the 28 titles contain approximatively 15.000 problems and our collection contains *only* 346). The real value of our book is the part which contains *all* the solutions of these exercises. We have tried to draw up these solutions as detailed as possible, so that each beginner can progress without skilled help.

The book is divided in two parts each consisting of seventeen chapters, the first part containing the exercises and the second part the solutions.

For the reader's convenience, each chapter begins with a short introduction giving the basic definitions and results one should know in order to solve the corresponding exercises. Some basic facts concerning groups and modules (vector spaces) are naturally assumed (e.g. cyclic groups, Lagrange theorem etc.). Also simple topology notions and results are assumed especially in the chapter devoted to rings of continuous functions.

A small part of this collection with hints and partial solutions, written by the first author, was ready in 1978, but only for internal use for problem sessions. The rest of the 346 exercises were almost all solved in problem sessions in the last 28 years. The chapter with exercises concerning rings of continuous functions is part of a one year course and series of seminars given by the second author.

Recently, the authors have become aware of the new publication *Exercises in Classical Ring Theory* written by T.Y. Lam, an excellent problem book published by Springer-Verlag. With only a few exceptions, this book actually does not contain *genuinely simple* exercises in ring theory, such as the exercises our collection provides. As Pham Ngoc Anh told us: "this is the book one should have before the problem book written by T.Y. Lam". Needless to say, the intersection of these two collections is nearly void. We therefore strongly hope that, using, in this order, these two books, one should have the best possible start in ring theory.

The first author acknowledges his colleague Horia F. Pop, lecturer in Computer Science, for his constant guidance and assistance in using computers and especially Latex and Scientific Word.

List of Symbols

Symbol	Description
$\mathbb{P}$	the set of the prime numbers
$\mathbb{N}$	the set of the non-negative integer numbers
$\mathbb{Z}$	the set of the integer numbers
$\mathbb{Q}$	the set of the rationals numbers
$\mathbb{R}$	the set of the real numbers
$\mathbb{C}$	the set of the complex numbers
$\mathcal{M}_n(R)$	the set of the $n \times n$-square matrices with entries in R
$E_{ij} = E_{ij}^a$	the matric units
S_n	the group of the permutations of degree n
$\mathbb{Z}_n$	the ring of the integers modulo n
$\mathbf{H}$	the ring of the quaternions
$\mathcal{P}(M)$	the set of the subsets of M
$\mathcal{P}_0(M)$	the set of the finite subsets of M
R^{op}	the opposite ring of R
$R[X]$	the ring of the polynomials of indeterminate X over R
$R\langle X \rangle = R[[X]]$	the ring of the power series of indeterminate X over R
$n(R)$	the number of the elements of the ring R
$Z(R)$	the center of R
$Id(R)$	the set of the idempotent elements of R
$U(R)$	the set of the units of R
$End(R)$	the ring of the endomorphisms of R
$Aut(K)$	the group of the automorphisms of K
$im(f)$	the image of f
$\ker(f)$	the kernel of f
$f^{-1}(Y)$	the preimage of Y by f
t_r	the left translation with r

(X)	the ideal generated by X
$l(X)$	the left annihilator of X
$r(X)$	the right annihilator of X
$R \times R'$	the direct product
$\prod_{i \in I} R_i$	the direct product
$S_1 \oplus S_2$	the direct sum
$char(R)$	the characteristics of R
$M \otimes_R N$	the tensor product of R-modules
$s(R)$	the socle of R
$rad(R)$	the (Jacobson) radical of R
$\mathcal{R}(R)$	the prime radical of R
$\mathcal{N}(R)$	the nilradical of R
$Spec(R)$	the prime spectrum
R_S	the ring of quotients
R_P	the localization of R in P
$Q(R)$	the classical (total) ring of quotients
τ^0	the discrete topology
τ_0	the indiscrete topology
$C(X)$	the continuous real-valued functions on X
$C^*(X)$	the bounded real-valued functions on X
$\underline{\alpha}$	the constant function with value α
$Z(f)$	the zero-set associated to f
$Z(X)$	the set of the zero-sets of the space X
χ_A	a characteristic function
$Sat_S(I)$	$= \{r \in R \mid \exists s \in S : rs \in I\}$
R^X	$= \{f \mid f : X \to R\}$
$A : B$	$= \{r \in R \mid \forall b \in B : rb \in A\}$
$\sqrt{I}$	$= \{r \in R \mid \exists n \in \mathbb{N} : r^n \in I\}$
$\mathbb{Q}^{(p)}$	$= \left\{\frac{m}{n} \mid (n; p) = 1\right\}$
$V(X)$	$= \{P \in Spec(R) \mid X \subseteq P\}$

Part I

EXERCISES

Chapter 1

Fundamentals

An element r of a ring R is called a

    left (right) zero divisor if there is a nonzero $s \in R : rs = 0(sr = 0)$;

    zero divisor if it is left and right divisor;

    left (right) cancellable if for every $a, b \in R : ra = rb(ar = br) \Rightarrow a = b$;

    idempotent if $r^2 = r$; two idempotents e, e' are *orthogonal* if $ee' = e'e = 0$; in a ring R we denote by $Id(R)$ the set of all the idempotent elements. A ring is called *Boole* if all its elements are idempotent.

    nilpotent if there is a $n \in \mathbb{N}^* : r^n = 0$;

    central if $rx = xr$ holds for each $x \in R$; central elements form a subring denoted by $Z(R)$ and called *the center*;

    regular if there is a $s \in R : r = rsr$.

If R has identity 1 an element r is called a

    left (right) unit if there is a $s \in R : rs = 1(sr = 1)$; s is called a *right (left) inverse* for r

    unit if it is left and right unit; the units form, together with multiplication, a group denoted $U(R)$.

    For two non-empty subsets $X, Y \subseteq R$ we denote by $X + Y = \{x + y | x \in X, y \in Y\}, X \cdot Y = \{xy | x \in X, y \in Y\}$ and

$$X \circ Y = \left\{ \sum_{i=1}^{n} x_i y_i | x_i \in X, y_i \in Y, 1 \leq i \leq n, n \in \mathbb{N} \right\}.$$

A subgroup S of $(R, +)$ is called a *subring* if $S \cdot S = \{ss' | s, s' \in S\} \subseteq S$. In a ring R with identity subrings may contain or not the identity.

3

If R and R' are two rings then $R \times R'$ together with componentwise addition and multiplication forms a ring called the *direct product* of R and R'. This generalises to an arbitrary family $\{R_i\}_{i \in I}$ of rings, the direct product being denoted by $\prod_{i \in I} R_i$. The canonical *projections* p_j : $\prod_{i \in I} R_i \to R_j, (j \in I)$ are defined by $p_j\left((r_i)_{i \in I}\right) = r_j$.

Universality of direct product: *for each family* $\{f_i : R \to R_i\}_{i \in I}$ *of ring homomorphisms there is a unique factorization homomorphism* $f : R \to \prod_{i \in I} R_i$ *such that* $f_i = p_i \circ f$.

A map $f : R \to R'$ between two rings is called a *ring homomorphism* if $f(x + y) = f(x) + f(y), f(x.y) = f(x).f(y)$. A bijective ring homomorphism is called *isomorphism*. A ring homomorphism between rings with identity is called *unital* if $f(1) = 1'$. We denote by $im(f) = f(R) = \{f(r)|r \in R\}$ the *image of* f and $\ker(f) = \{r \in R|f(r) = 0'\}$ the *kernel of* f.

We denote by $End(R)$ the set of all the group endomorphisms of $(R, +)$. Together with the usual addition and composition this forms a ring with identity 1_R.

We denote by $\mathcal{M}_n(R)$ the ring of the $n \times n$ (square) matrices with entries in the ring R together with the usual addition and multiplication.

The Dorroh extension *each ring can be embedded in a ring with identity.* In fact, if R is an arbitrary ring then $(R \times \mathbb{Z}, +, \cdot)$, together with $(r, n) + (s, m) = (r + s, n + m)$ and $(r, n) \cdot (s, m) = (rs + ns + mr, nm)$, is the stated extension $((0, 1)$ is the identity) and $\varphi : R \to R \times \mathbb{Z}, \varphi(r) = (r, 1)$ is the required embedding.

Ex. 1.1 In the definition of a ring with identity, the commutativity of the addition is a superfluous axiom.

Ex. 1.2 In each ring $(R, +, \cdot)$ we can define a new operation called the *circle composition* $a \circ b = a + b - ab, a, b \in R$.

(a) Show that $(R, \circ)$ is a monoid;

(b) Call an element $r \in R$ *right (left) quasi-regular* if there is an element $s \in R$ such that $r \circ s = 0$ $(s \circ r = 0)$. In a ring with identity show that the quasi-regular (right + left) elements form a group, with respect to the circle composition.

Ex. 1.3 An arbitrary ring R is commutative iff for each $a, b \in R$ one has $(a + b)^2 = a^2 + 2ab + b^2$.

Ex. 1.4 If a, b are elements in a ring R and $n, m \in \mathbb{N}^*, (n; m) = \gcd(n, m) = 1$ are such that $a^n = b^n$ and $a^m = b^m$ then $a = b$.

Ex. 1.5 On $\mathbb{R}^2$ consider the addition $(x_1, x_2) + (y_1, y_2) = (x_1 + y_1, x_2 + y_2)$ and two multiplications:

(a) $(x_1, x_2) \cdot (y_1, y_2) = (x_1 x_2, x_1 y_2 + x_2 y_1)$ and

(b) $(x_1, x_2) * (y_1, y_2) = (x_1 y_1 - x_2 y_2, x_1 y_2 + x_2 y_1)$.

Verify that $(\mathbb{R}^2, +, \cdot)$ and $(\mathbb{R}^2, +, *)$ are rings with identity. Find zero divisors in the first ring and prove that the second ring is isomorphic to $(\mathbb{C}, +, \cdot)$. Are these two rings isomorphic? Generalize.

Ex. 1.6 Let M be an arbitrary set and $\mathcal{P}(M)$ the set of all its subsets. Consider the operation $+ : \mathcal{P}(M) \times \mathcal{P}(M) \to \mathcal{P}(M)$ defined by $X + Y = (X \setminus Y) \cup (Y \setminus X)$ (also called the *symmetric difference*; also denoted by $\triangle$). Show that $(\mathcal{P}(M), +, \cap)$ is a commutative ring with identity and zero divisors, such that each element is idempotent.

Ex. 1.7 Describe all the ring structures on $(\mathbb{Z}, +)$.

Ex. 1.8 On $(\mathbb{Z}_n, +)$ all the ring with identity structures are isomorphic with the well-known ring $(\mathbb{Z}_n, +, \cdot)$.

Ex. 1.9 Describe all the nonisomorphic ring with identity structures one can define on a set with four elements. Hint: there are four such structures.

Ex. 1.10 Describe all the nonisomorphic noncommutative ring structures one can define on a set with four elements. Attention! Long solution.

Ex. 1.11 Let $p \in \mathbb{P}$. Show that there are only two nonisomorphic rings with p elements.

Ex. 1.12 Let $(R, +, \cdot)$ be a ring with identity and $(End(R, +), +, \circ)$ the corresponding endomorphism ring. Show that:
   (a) if for each $f \in End(R, +)$ and every $x \in R$ one has $f(x) = f(1).x$ then the following ring isomorphism holds

$$(R, +, \cdot) \cong (End(R, +), +, \circ)$$

   (b) If R is a commutative ring then $End(R, +)$ is commutative iff it is ring-isomorphic to R. Applications for $R = \mathbb{Z}$ and $R = \mathbb{Q}$.

Ex. 1.13 Let R be a ring with identity and $a \in R$.
   (a) If a has a left (or right) -sided inverse but no right (resp.left) -sided inverse then a has at least two left (or right) -sided inverses.
   (b) If a has more than a left (or right) -sided inverse then it has an infinite number of such inverses.

Ex. 1.14 Let $\mathcal{M}_n(R)$ denote the ring of all the $n \times n$-matrices with elements in a ring R ($n \in \mathbb{N}^*$).
   (a) Show that $\mathcal{M}_n(R)$ has identity iff R has identity.
   (b) For $n \geq 2$ show that $\mathcal{M}_n(R)$ is commutative iff $R^2 = \{0\}$ (i.e. $\forall a, b \in R : ab = 0$).
   (c) The center of $\mathcal{M}_n(R)$ consists of all the diagonal matrices $r.I_n$ such that $r \in Z(R) = \{r \in R | rx = xr, \forall x \in R\}$ (the center of R).

Ex. 1.15 If a ring R has no zero-square non-zero elements (i.e. $a \neq 0, a^2 = 0 \Rightarrow a = 0$) then prove that each idempotent element belongs to $Z(R)$ (the center of R).

Ex. 1.16 Show that if $r^2 - r \in Z(R)$ holds for every $r \in R$, then R is commutative.

Ex. 1.17 Compute the (multiplicative) group of the units (also called invertible elements) for the ring $\mathbb{Z}[i] = \{a + bi | a, b \in \mathbb{Z}\}$ of the Gauss integers.

Ex. 1.18 For a non-negative square-free integer d consider the ring $\mathbb{Z}[\sqrt{d}] = \{a + b\sqrt{d} | a, b \in \mathbb{Z}\}$; show that its group of units is infinite.

Ex. 1.19 Let R be a ring with identity and $a, b \in R$. If $1 - ab$ is a unit in R, show that $1 - ba$ is a unit in R too.

Ex. 1.20 The ring $\mathbb{Z}_n$ ($n \geq 2, n \in \mathbb{N}$) has non-zero nilpotent elements iff n is not square-free.

Ex. 1.21 Let $n = p_1^{r_1}..p_k^{r_k} \in \mathbb{N}^*$. Prove that $\overline{m} \in \mathbb{Z}_n$ is a nilpotent element iff $p_1..p_k$ divides m.

Ex. 1.22 Determine the idempotent elements of the ring $(\mathbb{Z}_n, +, .)$.

Application. The ring $\mathbb{Z}_n$ has no nontrivial idempotent elements iff n is a power of a prime number.

Ex. 1.23 (a) Prove that each Boole ring is commutative.

(b) (Faith) Show that a commutative ring is Boole iff it has no nonzero nilpotent elements and for each $a, b \in R : (a + b)ab = 0$.

Ex. 1.24 In a Boole ring show that for elements x, a, b if $a = xab$ then $a = xa$. Further, if $x = a + b + ab$ then $a = xa$ and $b = xb$. Finally, for $a_1, a_2, .., a_n$ elements find an element x such that $a_i = xa_i, \forall 1 \leq i \leq n$.

Ex. 1.25 Let x be a nilpotent element and y an unit in a commutative ring R with identity. Show that $x + y$ is an unit too. For $y = 1$, is one of the hypothesis unnecessary ? If the elements x, y do not commute show that the property does not hold.

Ex. 1.26 If $T = \{(a_{ij}) \in M_n(R) | a_{ij} = 0, \forall i, j \in \{1, 2, .., n\}, i > j\}$ is the set of all the triangular matrices, show that T is a subring of $M_n(R)$.

Ex. 1.27 Is $\{m + n\sqrt[3]{5} | m, n \in \mathbb{Z}\}$ a subring of $\mathbb{C}$?

Ex. 1.28 Show that $\mathbb{Z}_4 \times \mathbb{Z}_4$ has exactly three subrings with identity.

Ex. 1.29 Let $q \in \mathbb{Q}$ and $M = \left\{ \begin{pmatrix} a & b \\ qb & a \end{pmatrix} \mid a, b \in \mathbb{Q} \right\}$. Prove that M is a subring with identity of $\mathcal{M}_2(\mathbb{Q})$ which is a division ring iff there is no $r \in \mathbb{Q}$ such that $q = r^2$.

Ex. 1.30 Characterize all the subrings with identity of $(\mathbb{Q}, +, \cdot)$.

Ex. 1.31 In a ring R with identity such that $x^6 = x, \forall x \in R$ show that $x^2 = x, \forall x \in R$.

Chapter 2

Ideals

A subgroup I of $(R, +)$ is called *left (right) ideal* if

$R \cdot I = \{ri | r \in R, i \in I\} \subseteq I$ (resp. $I \cdot R \subseteq I$) and *ideal* if it is left and right ideal. We denote by $(X) = \cap \{I \text{ (left or right) ideal in } R | X \subseteq I\}$, if $X \subseteq R$, called the *(left or right) ideal generated by* X. In an arbitrary

ring $(X) = \{\sum_{i=1}^{n} r_i x_i + \sum_{k=1}^{m} x'_k r'_k + \sum_{s=1}^{l} r''_s x''_s r'''_s + \sum_{j=1}^{t} n_j y_j\}$

with $r_i, r'_k, r''_s, r'''_s \in R, x_i, x'_k, x''_s, y_j \in X, n_j \in \mathbb{Z}$ and the reader can provide the simplier forms if the ring has identity or is commutative (or both).

An ideal I is called *finitely generated (resp. principal)* if for a finite (resp. one element) subset F one has $I = (F)$. A domain R is called a *principal ideal domain* if all its ideals are principal.

An left ideal I of R is called

maximal if $I \neq R$ and it is not properly contained in any left ideal $\neq R$;

minimal if $I \neq (0)$ and it contains not properly any nonzero left ideal of R. Maximal, minimal right ideals and maximal, minimal ideals are defined in a similar way.

A subset $X \subseteq R$ is called

nil if each element in X is nilpotent;

nilpotent if there is a $n \in \mathbb{N} : X^n = X \circ X \circ .. \circ X = (0)$. The smallest n with this property is called *the nilpotency index* of X. Nil and nilpotent left (right) ideals or ideals are defined in a similar way.

In a ring with identity each proper (left,right) ideal is contained in a maximal (left,right) ideal.

A ring R is called *regular (von Neumann)* if for each element $a \in R$ there is an element $b \in R : a = aba$.

Noether Isomorphism Theorems. (1) *If $f : R \to R'$ is a ring homomorphism then*

$$f(R) \cong R/\ker(f).$$

(2) *If I, J are ideals in a ring R then $(I + J)/I \cong J/(I \cap J)$.*

(3) *If $I \subseteq J$ are ideals in a ring R then J/I is an ideal in R/I and $(R/I)/(J/I) \cong R/J$.*

A right ideal I of a ring R is called *modular* if there is an element $e \in R$ such that for all $r \in R$ the element $r - er \in I$ (R has a left identity element modulo I).

The subset $l(X) = Ann.l.(X) = \{r \in R | rx = 0, \forall x \in X\}$ for a subset X of a ring R is called the *left annihilator* of X in R. Similarly,

$r(X) = Ann.r.(X) = \{r \in R | xr = 0, \forall x \in X\}$ is called the *right annihilator of X in R.*

Ex. 2.1 Let $F = \{f | f : [-1, 1] \to \mathbb{R}\}$ the commutative ring together with the usual addition and multiplication. Which of the following subsets are subrings and which are ideals:

(i) P the polynomial functions;

(ii) P_n the polynomial functions of degree at most n $(n \in \mathbb{N}^*)$;

(iii) Q_n the polynomial functions of degree n;

(iv) $A = \{f \in F | f(0) = 0\}$;

(v) $B = \{f \in F | f(0) = 1\}$.

Ex. 2.2 Let $R = \mathcal{P}(T)$, where $T = [0, 1] \subseteq \mathbb{R}$, the ring together with the usual ring laws of symmetric difference and intersection (see 1.6). For $A = \left[0, \frac{1}{2}\right]$ and $B = \left\{\frac{1}{4}\right\}$ compute $(A), (B), (A, B)$ and $(A) \circ (B)$, the ideals generated by the corresponding subsets of T.

Ex. 2.3 (a) For a subset X of a ring R show that $l(X)$ is a left ideal and $r(X)$ is a right ideal in R.

(b) If X is a left (right) ideal in R then $l(X)$ (resp. $r(X)$) is an ideal in R.

(c) The following inclusions and equalities hold: (i) $X \subseteq Y \Rightarrow l(Y) \subseteq l(X)$ and $r(Y) \subseteq r(X)$; (ii) $X \subseteq r(l(X))$ and $X \subseteq l(r(X))$; (iii) $l(X) = l(r(l(X)))$ and $r(X) = r(l(r(X)))$.

Ex. 2.4 Show that if $R = I + J$ holds with I, J right modular ideals then $I \cap J$ is a right modular ideal too.

Ex. 2.5 Find all the ideals of $(n\mathbb{Z}, +, \cdot)$ for $n \in \mathbb{N}^*$.

Ex. 2.6 Let I, J, K be ideals in a ring R. Show that $I \subseteq K$ implies $(I + J) \cap K = I + (J \cap K)$.

Ex. 2.7 In $\mathbb{Z}$ show that the set-reunion of two ideals needs not to be an ideal too.

Ex. 2.8 Give an example of ring (without identity) such that not every ideal is included in a maximal ideal.

Ex. 2.9 If M is a maximal right ideal of a ring R with identity and $a \in R \backslash M$ then verify that $a^{-1}M = \{r \in R | ar \in M\}$ is also a maximal right ideal.

Ex. 2.10 Show that in a Boole ring each finitely generated ideal is principal.

Ex. 2.11 For two commutative rings R and S, determine the ideals of the direct product (sum) $R \times S$. Applications: $\mathbb{Z} \times \mathbb{Z}$ and $K \times K$ for a division ring K.

Ex. 2.12 Show that the following implication is false: "if A is a left ideal and B is a right ideal in the same ring R then $A \cap B$ is an ideal of R ".

Ex. 2.13 Let $A : B$ denote $\{r \in R | \forall b \in B : rb \in A\}$ for A and B ideals of a ring R.

(a) Show that $A : B$ is an ideal of R;

(b) $A : B$ is the l.u.b.$\{I \trianglelefteq R | IB \subseteq A\}$.

(c) Verify the following equalities : $A : A = R$; $(A_1 \cap .. \cap A_n) : B = (A_1 : B) \cap .. \cap (A_n : B)$ and $A : (B_1 + .. + B) = (A : B_1) \cap .. \cap (A : B_n)$.

(d) In $\mathbb{Z}$ show that $n\mathbb{Z} : m\mathbb{Z} = \frac{[n;m]}{m}\mathbb{Z}$ (here $[n;m]$ denotes the l.c.m.(n,m)).

Ex. 2.14 In a commutative ring R prove the following properties:

(i) $(A : B) \circ B \subseteq A, (A : (A + B)) = (A : B)$;

(ii) $((A : B) : C) = (A : (B \circ C)) = (A : (B : C))$;

(iii) if R has identity $A : B = R$ iff $B \subseteq A$.

Ex. 2.15 Let I be an ideal in a commutative ring R.

(a) Show that $\sqrt{I} = \{r \in R | \exists n \in \mathbb{N} : r^n \in I\}$ is an ideal too.

(b) Verify the following equalities: $\sqrt{\sqrt{I}} = \sqrt{I}$; $\sqrt{I \cap J} = \sqrt{I} \cap \sqrt{J}$ and $\sqrt{I + J} = \sqrt{\sqrt{I} + \sqrt{J}}$.

(c) Can this exercise be used in order to show that the set of all the nilpotent elements form in a commutative ring an ideal ? Is commutativity essential ?

Ex. 2.16 Let I be a left ideal and J an ideal of R. If I, J are nil (or nilpotent) ideals show that $I + J$ has the same property.

Ex. 2.17 Verify the following properties: subrings and quotient rings of nil rings are nil; for an ideal I in R, if I and R/I are nil then R is nil.

Ex. 2.18 For a prime number p and $n \in \mathbb{N}, n > 1$ show that each proper ideal of $\mathbb{Z}_{p^n}$ is nilpotent. For each n let I_n be a proper ideal in $\mathbb{Z}_{p^n}$ and

$$I = \left\{ (x_n) \in \prod_{n \in \mathbb{N}^*} \mathbb{Z}_{p^n} | x_n \in I_n \text{ and } (x_n) \text{ has finite support} \right\}.$$

Show that I is a nilideal but is not a nilpotent ideal.

Ex. 2.19 Let U be an ideal of the ring R. We say that in R idempotents can be *lifted* modulo U if for each idempotent element $y \in R/U$ there is an idempotent $x \in R$ such that $x + U = y$. Show that if U is a nilideal of R then the idempotents can be lifted modulo U.

Ex. 2.20 Let R be a ring with identity.
  (a) Show that every ideal in the ring of all square matrices $M_n(R)$ has the form $M_n(A)$ where A is an ideal of R;
  (b) Verify the ring isomorphism $M_n(R)/M_n(A) \cong M_n(R/A)$.
  (c) In $M_2(2\mathbb{Z})$, using the set $S = \{a_{ij} \in M_2(2\mathbb{Z}) | a_{11} \in 4\mathbb{Z}\}$ show that the existence of the identity of the ring is essential.
  (d) The result from (a) holds for left (or right) ideals ?

Ex. 2.21 Show that $M_2(\mathbb{R})$ has no nontrivial ideals.

Ex. 2.22 Give an example of a non-commutative ring with a proper commutative quotient ring.

Ex. 2.23 Let X be a non-empty set and Y a proper subset of X. Consider $\mathcal{P}(X)$ and $\mathcal{P}(X \backslash Y)$ as (boolean) rings (see 1.6) relative to the symmetric difference and the intersection. Show that:
  (a) $\mathcal{P}(X \backslash Y) \cong \mathcal{P}(X)/\mathcal{P}(Y)$;
  (b) If X is finite every ideal of $\mathcal{P}(X)$ has the form $\mathcal{P}(Y)$ for a suitable subset Y of X;
  (c) If X is infinite (b) fails.

Ex. 2.24 Let I and J be ideals in a ring R. Prove that the canonical ring homomorphism $R \ / \ I \cap J \to R/I \times R/J$ is an isomorphism iff $I + J = R$ (so called *comaximal* ideals).

Ex. 2.25 Let R be a commutative ring and $I = I^2$ a finitely generated ideal of R. Find an idempotent element $e \in R$ such that $I = Re$.

Ex. 2.26 Show that in a commutative regular (von Neumann) ring every finitely generated ideal is principal.

Ex. 2.27 For a prime number p let $\mathbb{Q}^{(p)} = \left\{ \frac{m}{n} \in \mathbb{Q} | (n; p) = 1 \right\}$ (as above $(n; p)$ denotes the g.c.d.(n, p) and all fractions are irreducible). Verify the following properties:

(a) $\mathbb{Q}^{(p)}$ is a subring of $\mathbb{Q}$;

(b) For every $x \in \mathbb{Q}$ either $x \in \mathbb{Q}^{(p)}$ or $x^{-1} \in \mathbb{Q}^{(p)}$;

(c) The only subrings of $\mathbb{Q}$ which contain $\mathbb{Q}^{(p)}$ are $\mathbb{Q}^{(p)}$ and $\mathbb{Q}$;

(d) Every ideal of $\mathbb{Q}^{(p)}$ has the form $(p^n) = p^n \mathbb{Q}^{(p)}$ for a suitable $n \in \mathbb{N}$;

(e) $\bigcap_{p \in \mathbb{P}} \mathbb{Q}^{(p)} = \mathbb{Z}$ (here $\mathbb{P}$ denotes the set of all the prime numbers).

Chapter 3

Zero Divisors

Zero divisors are defined in the introduction of the first chapter.

A nonvoid subset S of a ring R is called a *multiplicative system* if it is closed under multiplication.

A proper ideal P is called *prime* if for each ideals I, J the inclusion $I \circ J \subseteq P$ implies $I \subseteq P$ or $J \subseteq P$.

Ex. 3.1 In a ring with identity show that each idempotent element which differs from 0 and 1 is a zero divisor.

Ex. 3.2 Let $R = \mathcal{C}[0,1] = \{f : [0,1] \to \mathbb{R} | f \text{ continuous}\}$ be ring with the usual addition and multiplication. Prove that $f \in R$ is a zero divisor iff there is an open interval in $[0,1]$ such that the restriction of f vanishes. Find also the idempotent resp. nilpotent elements from R.

Ex. 3.3 Let M be an infinite arbitrary set, $(\mathcal{P}(M), +, \cap)$ the ring of all the subsets of M (see 1.6) and $\mathcal{P}_0(M)$ the subset of all the finite subsets of M. Show that:
  (i) $\mathcal{P}_0(M)$ is a subring of $\mathcal{P}(M)$;
  (ii) each element from $\mathcal{P}_0(M)$ is a zero divisor in $\mathcal{P}_0(M)$;
  (iii) each element $\neq M$ from $\mathcal{P}(M)$ is a zero divisor in $\mathcal{P}(M)$.

Ex. 3.4 Let $m, n \in \mathbb{N}^*, \gcd(m; n) = 1$ and $q = mn$. If
  $T = \{\overline{mx} | x \in \mathbb{Z}, (x; n) = 1\} \subseteq \mathbb{Z}_q$, show that T is closed under multiplication in $\mathbb{Z}_q$ and $(T, \cdot)$ is a group of zero divisors from $\mathbb{Z}_q$.

Ex. 3.5 Show that a finite ring R with left and right non-zero divisors has identity.

Ex. 3.6 Let S be a multiplicative system in a commutative ring R which contains only non-zero divisors. Show that the following subset $T = \{a \in R | \exists s \in S, \exists r \in R : s = ar\}$, has the same properties.

Ex. 3.7 Let $(0) \neq I$ an ideal without non-zero divisors in a ring R. If there are elements $0 \neq a \in I$ and $r \in R$ such that $ra = 0$ then $rI = Ir = (0)$.

Ex. 3.8 If I is an ideal of the ring R then show that if R has no non-zero divisors then $R/(l(I) \cap r(I))$ has the same property.

Ex. 3.9 Prove that every ring without non-zero divisors can be embedded in a ring with identity which has no non-zero divisors.

Ex. 3.10 Let R be a commutative ring and $f \in R[X]$. Prove that f is a zero divisor in $R[X]$ iff there is an element $0 \neq r \in R$ such that $rf = 0$.

Ex. 3.11 If R is a finite ring with identity then each non-zero element is either a one-sided zero divisor or a unit in R.

Ex. 3.12 Let K be a field and $A \in \mathcal{M}_2(K), A \neq 0$. Show that A is a unit iff A is a left non-zero divisor in $\mathcal{M}_2(K)$. If K is not a field, does this hold ?

Ex. 3.13 If $M = \left\{ \begin{pmatrix} a+bi & c+di \\ -c+di & a-bi \end{pmatrix} \in \mathcal{M}_2(\mathbb{C}) \,|\, a, b, c, d \in \mathbb{R} \right\}$,

verify that M is a ring without zero divisors.

Ex. 3.14 Let R be a ring, $a \in R$ and $0 \neq b \in R$ such that $aba = 0$. Prove that a is left or right zero divisor in R.

Ex. 3.15 In a commutative regular (von Neumann) ring show that each nonunit is a zero divisor.

Ex. 3.16 Do the zero divisors of a ring R form an ideal ?

Ex. 3.17 Show that the set D of all the zero divisors of a commutative ring contains at least one prime ideal.

Ex. 3.18 If D is an integral (commutative) domain show that $D\langle X \rangle = D[[X]]$, the ring of all the power series with coefficients in D is also an integral domain.

Ex. 3.19 For a surjective ring homomorphism $f : R \to R'$ of commutative rings with identity analyse the following statement: "$f(r)$ is a zero divisor iff r is a zero divisor".

Chapter 4

Ring Homomorphisms

Let $f : R \to R'$ a ring homomorphism. If the rings have identity f is called *unital* if $f(1) = 1'$.

$\ker(f) = \{r \in R | f(r) = 0'\}$ is an ideal of R.

For any subset $Y \subseteq R'$ one defines the *preimage*

$f^{-1}(Y) = \{r \in R | f(r) \in Y\}$.

Correspondence theorem. *If A is a subring and I an ideal of R then $f(A)$ is a subring in R and $f(I)$ is an ideal in $f(R)$. Conversely, if A' is a subring and I' an ideal of R' then $f^{-1}(A')$ is a subring and $f^{-1}(I')$ is an ideal of R. If f is surjective, $F : \{A$ subring in $R| \ker(f) \subseteq A\} \to \{$ subrings of $R'\}$ defined by $F(A) = f(A)$ is an order isomorphism which preserves and reflects ideals and in particular maximal ideals.*

If R, R' are rings there is always at least one ring homomorphism: $o : R \to R', o(r) = 0', \forall r \in R$ the *zero (trivial) homomorphism*.

Let a be a fixed element in a ring R. Define the map $t_a : R \to R$ by $t_a(r) = ar, \forall r \in R$, the *left translation* with a (right translations are defined in a symmetrical way). The translations are group endomorphisms of $(R, +)$.

Ex. 4.1 Show that for a homomorphism $f : R \to R'$ of rings with identity, $f(1)$ is an identity in $f(R)$ but not necessarily in R'.

   Application: each surjective homomorphism of rings with identity is unital.

Ex. 4.2 Let K be a simple ring and $f : K \to R$ a ring homomorphism. Prove that f is either trivial or injective.

Ex. 4.3 Let R be a ring and $f, g : \mathbb{Q} \to R$ be ring homomorphisms such that $f(n) = g(n), \forall n \in \mathbb{Z}$. Prove that $f = g$.

Ex. 4.4 Prove that the map $f : \mathbb{Z} \to \mathbb{Z}_3 \times \mathbb{Z}_5, f(x) = ([x]_3, [x]_5)$ is a surjective ring homomorphism and determine its kernel.

Ex. 4.5 If, for a non-zero square-free integer d, $\mathbb{Z}\left[\sqrt{d}\right]$ denotes the subring $\left\{a + b\sqrt{d} | a, b \in \mathbb{Z}\right\}$ of $\mathbb{C}$, find all the elements x such that the composition $\mathbb{Z} \to \mathbb{Z}\left[\sqrt{d}\right] \to \mathbb{Z}\left[\sqrt{d}\right]/(x)$ is surjective. As a special case, when is the canonical map $\mathbb{Z}/(x) \cap \mathbb{Z} \to \mathbb{Z}\left[\sqrt{d}\right]/(x)$ a ring isomorphism ?

Ex. 4.6 Give an example of a ring homomorphism $f : R \to R'$ and an ideal I of R such that $f(I)$ is not an ideal of R'.

Ex. 4.7 Determine (a) all the unital ring endomorphisms of $\mathbb{Z}$ and (b) all the ring homomorphisms from $\mathbb{Z}$ to $\mathbb{Q}$.

Ex. 4.8 For $n \geq 2, n \in \mathbb{N}$ determine the ring endomorphisms of $\mathbb{Z}_n$.

Ex. 4.9 For $n, m \in \mathbb{Z}, n, m \geq 2$ determine the ring homomorphisms $\mathbb{Z}_n \to \mathbb{Z}_m$.

Ex. 4.10 Let $f : R \to R'$ be a surjective homomorphism of rings with identity and $a \in R$. Show that if a is either idempotent, nilpotent, central or a unit then $f(a)$ is also idempotent, nilpotent, central respectively a unit. Is any converse true ?

Ex. 4.11 Let $f : R \to R'$ be an isomorphism of rings with identity.

(a) Show that f preserves elements which satisfy the relation $x^2 = 1 + 1$;

(b) Infer : $\mathbb{Z}[\sqrt{2}]$ is not isomorphic to $\mathbb{Z}[\sqrt{3}]$. Generalize.

Ex. 4.12 Prove that $R = \left\{ \begin{pmatrix} 0 & a \\ 0 & b \end{pmatrix} \mid a, b \in \mathbb{Q} \right\}$ is a noncommutative subring of $\mathcal{M}_2(\mathbb{Q})$. If $I = \{A \in R \mid A^2 = 0\}$ show that I is an ideal and $R/I \cong \mathbb{Q}$.

Ex. 4.13 Let $f : R \to R'$ be a ring homomorphism, A an ideal in R and $\{B_i\}_{i \in I}$ ideals in R'. Show that $f^{-1}(f(A)) = A + \ker(f)$ and
$$f^{-1}\left(\bigcap_{i \in I} B_i \right) = \bigcap_{i \in I} f^{-1}(B_i).$$

Ex. 4.14 Give an example of a homomorphism $f : R \to R'$ of commutative rings and a maximal ideal M in R' such that $f^{-1}(M) \neq R$ but $f^{-1}(M)$ is not a maximal ideal.

Ex. 4.15 Does a commutative ring with identity R such that $R[X] \cong (\mathbb{Z}, +, \cdot)$, exist ?

Chapter 5

Characteristics

Denote by $ord(a) = ord_{(R,+)}(a)$ the (group) order of an element a in a ring R. If there is a $m \in \mathbb{N}^*$ such that $ma = 0, \forall a \in R$ then we denote by $char(R)$ the smallest positive integer (if it exists) having this property (i.e. all elements are of finite order and $ord(a)$ divides $char(R)$). In the remaining case (i.e. there are elements of infinite order or $\{ord(a)|a \in R\}$ is not bounded) we say that $char(R) = 0$.

If a ring R has identity then $char(R) = ord_{(R,+)}(1)$. More generally, if a is a unit of R then $char(R) = ord_{(R,+)}(a)$.

Each ring with identity without zero divisors (e.g. integral domains, division rings, fields) has prime or zero characteristics.

Ex. 5.1 Show that the characteristic of a Boole ring (1.23) is 2.

Ex. 5.2 Compute the characteristic of $\mathbb{Z} \times \mathbb{Z}, \mathbb{Z}_3 \times \mathbb{Z}, End(\mathbb{Z})$ and $End(\mathbb{Z}_3)$.

Ex. 5.3 For $m, n \in \mathbb{N}^*$ prove that the characteristic of $\mathbb{Z}_m \times \mathbb{Z}_n$ is $[m; n]$, the least common multiple of m and n. Generalize.

Ex. 5.4 Compute the characteristic of $\mathbb{Q} \times \mathbb{Z}_8$.

Ex. 5.5 Show that each division ring has a natural structure of vectorial space over $\mathbb{Z}_p$ (for a prime number p) or $\mathbb{Q}$.

Ex. 5.6 Prove that for each ring R with identity there is a unique unital ring homomorphism $f : \mathbb{Z} \to R$. Moreover, if $char(R) = 0$ then f is injective and if $char(R) = n \in \mathbb{N}^*$ then $\ker(f) = (n) = n\mathbb{Z}$.

Ex. 5.7 If there is an unital ring homomorphism $f : K \to K'$ between two division rings then prove that $char(K) = char(K')$.

Ex. 5.8 For two fields K, K' of characteristics different from 2 and 3 and a map $f : K \to K'$ such that $f(x + y) = f(x) + f(y), \forall x, y \in K$ and $f(1) = 1'$ show that f is a (unital) ring homomorphism iff $f(x^3) = f(x.)^3, \forall x \in K$.

Ex. 5.9 Give an example of commutative ring with identity of prime characteristic which is not a field.

Ex. 5.10 Show that the characteristic of K_8 (see 7.2) is 2.

Ex. 5.11 Let us define on $\mathbb{Z}_3 \times \mathbb{Z}_3$ the following laws: $(a, b) + (c, d) = (a+c, b+d)$ and $(a, b).(c, d) = (ac - bd, ad + bc)$. Verify that we obtain a field (further denoted K_9, see also 7.3) and compute its characteristic.

Ex. 5.12 Give an example of an infinite division ring of finite (and hence prime) characteristic.

Ex. 5.13 Let K be a field of characteristic 0. If $A, B \in M_n(K)$ is $AB - BA = I_n$ possible ?

Ex. 5.14 Let K be a field and $char(K) = p \neq 0$. Prove that the polynomial $X^{p^n} - a \in K[X]$, for $a \in K$, has at most a root in K.

Ex. 5.15 Let K be a division ring and $char(K) \neq 2$. Show that $X^2 + Y^2 - 1$ is irreducible in $K[X, Y]$.

Ex. 5.16 Let K be a field of characteristic p and a, b non-zero elements of K. Prove that $f \in K[X]$ is either irreducible or factors completely in $K[X]$ for:

(i) $f = X^p - X - a$;
(ii) $f = X^p - b^{p-1}X - a$;
(iii) $f = X^p + b^{p-1}a^{-1}X - a^{-1}$.

Ex. 5.17 Let K be a field such that $char(K) \neq 2$ and $a \in K$ such that $a^{2n+1} = 1$ for $n \in \mathbb{N}^*$. Find a map $f : K \to K$ such that $f(x) + f(ax) = 2x, \forall x \in K$ (here $2 = 1 + 1$ in K).

Chapter 6

Divisibility in Integral Domains

If $a, b \in R$ a commutative ring, we say that b *divides* a (denoted $b|a$) if there is an element $c \in R : a = bc$. One checks $b|a \Leftrightarrow aR = (a) \subseteq (b) = bR$. Two elements are *associated (in divisibility)* denoted by $a \sim b$ if $b|a$ and $a|b$. *This is an equivalence relation and* $a \sim b \Leftrightarrow aR = bR$. An element p is called

irreducible if it is not a unit and has no other divisors then units and associated elements;

reducible if it is not irreducible;

prime if it is not a unit and $p|a.b \Rightarrow p|a$ or $p|b$;

greatest common divisor for $r_1, r_2, .., r_n$ if $p|r_i (1 \leq i \leq n)$ and if $d|r_i (1 \leq i \leq n)$ then $d|p$;

least common multiple is defined similarly;

Each prime element is irreducible.

An integral domain D is called *Gauss* or a *unique factorization domain* if each $a \in D$ which is not a unit decomposes in a product of irreducible elements, each two such decompositions being associated.

In each Gauss ring each two elements have g.c.d. and each irreducible element is prime.

Each principal ideal domain is a Gauss ring.

A pair (D, δ) which consists of an integral domain D and a map $\delta : D^* \rightarrow \mathbb{N}^*$ is called an *Euclidean ring* if $\forall a \in D, \forall b \in D^*, \exists q, r \in D : a = bq + r$ where $r = 0$ or $\delta(r) < \delta(b)$.

Each Euclidean ring is a principal ideal domain (and hence, a Gauss ring).

Ex. 6.1 Let $D \neq 0$ be an integral domain such that each pair of nonzero elements from D are associated. Prove that D is a field.

Ex. 6.2 A commutative ring with identity R is called *normed* if there is a map $N : R \to \mathbb{N}$ such that $N(xy) = N(x)N(y)$ (also called a *multiplicative* map), $N(x) = 0$ iff x divides 0 and $N(x) = 1$ iff x is a unit in R. In such a ring show that every element decomposes into irreducible elements.

Ex. 6.3 Show that each Gaussian ring is normed.

Ex. 6.4 Verify that the rings $\mathbb{Z}[\sqrt{n}]$ and $\mathbb{Z}[i\sqrt{n}]$ where $n \in \mathbb{N}, n > 1$ is not a square in $\mathbb{Z}$,are normed.

Ex. 6.5 Show that $\mathbb{Z}[i]$ is an Euclidean ring.

Ex. 6.6 Decompose the following numbers in $\mathbb{Z}[i\sqrt{2}] : 5, 1 + i\sqrt{2}, 2 + i\sqrt{2}$.

Ex. 6.7 In $\mathbb{Z}[i]$, the ring of Gauss integers, find a gcd for 5 and $1 + 3i$.

Ex. 6.8 Show that in $\mathbb{Z}[i\sqrt{5}]$ the elements 6 and $2(1 + i\sqrt{5})$ don't have a gcd .

Ex. 6.9 Find a gcd for 3 and $1 + i\sqrt{5}$ in $\mathbb{Z}[i\sqrt{5}]$.

Ex. 6.10 Using an Euclidean algorithm find in $\mathbb{Z}[i]$ a generator for the ideal $(1 + i, 4 + i)$.

Ex. 6.11 Let $n \in \mathbb{N}^*, n \geq 3$ such that $\sqrt{n} \notin \mathbb{Z}$. Show that in the ring $\mathbb{Z}[\sqrt{n}]$ the element 2 is irreducible but not prime.

Ex. 6.12 In $\mathbb{Z}[i]$ prove that $1 + 2i$ is a prime element.

Ex. 6.13 In the ring $R = \mathbb{Z}[i\sqrt{6}]$ verify that the following hold:
    (a) there are no elements of norm equal to 2 or 5;
    (b) the elements $2, 5, 2 - i\sqrt{6}$ are irreducible but not prime;
    (c) in R the identity 1 is a $\gcd(5, 2 + i\sqrt{6})$ but $1 \notin (5, 2 + i\sqrt{6})$.
    (d) a $\gcd(10, 4 + 2i\sqrt{6})$ does not exist;
    (e) in R the element 10 can be decomposed into two different products of irreducible non-associated elements.

Ex. 6.14 Show that there are an infinite number of irreducible elements in $\mathbb{Z}[i]$.

Ex. 6.15 Let $D \neq 0$ be an integral domain such that each element from D^* is a unit or an irreducible element in D. Prove that D is a field.

Ex. 6.16 Show that $\mathbb{Z}[i\sqrt{3}]$ and $\mathbb{Z}[i\sqrt{6}]$ are not Gaussian rings.

Ex. 6.17 Is the ring $\mathbb{Z}[i\sqrt{2}]$ Euclidean ?

Ex. 6.18 Verify that the ring $\mathbb{Z}[\sqrt{3}]$ is Euclidean.

Chapter 7

Division Rings

A ring K with identity is called a *division ring* if $(K^*, \cdot)$ is a group. A commutative division ring is called a *field*.

Each division ring has zero or prime characteristics.

Wedderburn's Theorem. *Every finite division ring is commutative (a field).*

The number of elements in a finite field is a power of its characteristics (see 7.15).

Any two finite fields with the same number of elements are isomorphic.

Ex. 7.1 Show directly that each field with 4 elements is isomorphic with K_4 (see 1.9).

Ex. 7.2 On $\mathbb{Z}_2^3$ define the following 2 laws: $(a, b, c) + (u, v, w) = (a + u, b + v, c + w)$ and $(a, b, c) \cdot (u, v, w) = (au + bw + cv, av + bu + bw + cv + cw, aw + bv + cu + cw)$. Show that a field (further denoted K_8) is defined in this way.

Ex. 7.3 On $\mathbb{Z}_3^2$ define the following 2 laws:
$(a, b) + (c, d) = (a + c, b + d)$ and
$(a, b) \cdot (c, d) = (ac - bd, ad + bc), \forall a, b, c, d \in \mathbb{Z}_3$.
Verify that in this way one obtains a field with 9 elements, denoted K_9.

Ex. 7.4 Show that $\left\{ \begin{pmatrix} a & b \\ -b & a \end{pmatrix} \mid a, b \in \mathbb{R} \right\} \subseteq M_2(\mathbb{R})$ together with the usual matrices addition and multiplication forms a field isomorphic to $\mathbb{C}$.

Ex. 7.5 Show that $\left\{ \begin{pmatrix} x + 2y & 3y \\ 2y & x - 2y \end{pmatrix} \mid x, y \in \mathbb{Q} \right\} \subseteq M_2(\mathbb{Q})$ together with the usual matrices addition and multiplication forms a field isomorphic with $\mathbb{Q}[\sqrt{10}]$.

Ex. 7.6 Let d be a square-free integer. Show that $\left\{ \begin{pmatrix} a & b \\ db & a \end{pmatrix} \mid a, b \in \mathbb{Q} \right\} \subseteq M_2(\mathbb{Q})$ together with the usual matrices addition and multiplication forms a field isomorphic to $\mathbb{Q}[\sqrt{d}]$.

Ex. 7.7 Let K be a field. Prove that the groups $(K, +)$ and $(K^*, \cdot)$ are not isomorphic.

Ex. 7.8 Let K be a field and $char(K) \neq 0$. Prove that there is only one group homomorphism $(K, +) \to (K^*, \cdot)$. Can this be applied to $\mathbb{Q}$?

Ex. 7.9 Let D be a noncommutative division ring. If there are $a, b \in D, n \in \mathbb{Z}$ such that $a^n b^n - b^{n+1} a^{n+1} = 1$ and $a^{2n+1} + b^{2n+1} = 0$ then show that $b^n a^n - a^{n+1} b^{n+1} = 1$.

Ex. 7.10 A non-zero ring R is a division ring iff each $x \in R, x \neq 0$ has a right inverse.

Ex. 7.11 A non-zero ring R is a division ring iff for each $x \in R., x \neq 1$ one has $x + y = xy$ for a suitable $y \in R$.

Ex. 7.12 Give an example of an infinite non-commutative division ring.

Ex. 7.13 Prove the existence of a field with p^2 elements for each prime number p.

Ex. 7.14 Compute the center of our solution to the previous exercise.

Ex. 7.15 Show that each finite division ring K has as number of elements a power of a prime number.

Ex. 7.16 Let K be a field and f, g nonzero endomorphisms of K. Show that $H = \{x \in K | f(x) = g(x)\}$ is a subfield of K.

Ex. 7.17 If K and K' are division rings and for $n, m \in \mathbb{N}^*$ an isomorphism $\mathcal{M}_n(K) \cong \mathcal{M}_m(K')$ holds, then $n = m$ and $K \cong K'$.

Chapter 8

Automorphisms

A ring homomorphism $f : R \to R'$ is called *automorphism* if it is bijective (an isomorphism) and $R = R'$ (an endomorphism).

Each automorphism is unital.

Under usual composition and inverse *the automorphisms of a field K form a group, denoted Aut(K).*

Ex. 8.1 Prove that the only automorphism of $\mathbb{Q}$ is the identical one.

Ex. 8.2 Determine all the automorphisms of the field $(\mathbb{R}, +, .)$ and the automorphisms of the field $(\mathbb{C}, +, .)$ for which $\mathbb{R}$ is invariant.

Ex. 8.3 Describe the automorphisms of $\mathbb{Q}[\sqrt{d}]$ for a square-free non-negative integer d.

Ex. 8.4 Determine the automorphisms of K_4 (see 1.9).

Ex. 8.5 Determine the automorphisms of K_8 (see 7.2).

Ex. 8.6 Show that K_9 (see 7.3) has only two automorphisms.

Ex. 8.7 Give examples of non-identical automorphisms of $\mathbb{Z}[\sqrt{3}]$.

Ex. 8.8 Determine the automorphisms of $\mathbb{Z}_p$.

Ex. 8.9 Let R be an integral domain.
(a) If $a, b \in R$ and a is a unit, the map $\varphi : R[X] \to R[X]$ defined by $\varphi(f(X)) = f(aX + b)$ is an automorphism of $R[X]$.
(b) Prove that each automorphism of $R[X]$ whose restriction to R is the identical automorphism has the form pointed out in (a);
(c) Determine the automorphisms of $\mathbb{Z}_p[X]$ (for p a prime number) and of $\mathbb{Q}[X]$.

Ex. 8.10 Prove that $\mathbb{Q}(\sqrt[3]{a})$ has only one automorphism, for each $a \in \mathbb{Q}_+$ which is not a cube in $\mathbb{Q}$.

Ex. 8.11 Let K be a (commutative) finite field of prime characteristic p. Show that the map $f : K \to K, f(x) = x^p, \forall x \in K$ is an automorphism of K (also called the *Frobenius* automorphism).

Ex. 8.12 Prove that each division ring which has only a finite number of automorphisms is commutative.

Ex. 8.13 For a finite ring R with identity denote by $n(R)$ the number of elements of R and by $a(R)$ the number of ring automorphisms of R. Determine (up to a ring isomorphism) the rings R such that $n(R) \geq 3$ and $a(R) = (n(R) - 2)!$ (long solution !).

Chapter 9

The Tensor Product

Let M_R be a right R-module, $_RN$ a left R-module and G an abelian group. A $\mathbb{Z}$-bilinear map $\varphi : M \times N \to G$ is called R-balanced if $\varphi(mr, n) = \varphi(m, rn)$ holds for each $m \in M, n \in N, r \in R$.

We denote by L the free $\mathbb{Z}$-module over $M \times N$ i.e. $L = \mathbb{Z}^{(M \times N)}$ and by $\{[m, n] \,|\, (m, n) \in M \times N\}$ its canonical basis. Let K be the subgroup of L generated by the elements of the form $[m_1 + m_2, n] - [m_1, n] - [m_2, n]$

$[m, n_1 + n_2] - [m, n_1] - [m, n_2]$ and $[mr, n] - [m, rn]$ for $m, m_1, m_2 \in M, n, n_1, n_2 \in N, r \in R$. We denote by $M \otimes_R N = L/K$ and by $\varphi : M \times N \to M \otimes_R N, \varphi((m, n)) = [m, n] + K \overset{not}{=} m \otimes n$ and call this the tensor product of M and N. Every element in $M \otimes_R N$ can be written as a finite sum $\sum_{i=1}^{k} m_i \otimes n_i$.

The tensor product is an abelian group and φ is a $\mathbb{Z}$-bilinear R-balaced map.

If $_SM_R$ and $_RN_T$ are bimodules the tensor product has a natural structure of $S - T$-bimodule.

The universality of the tensor product. For each $\mathbb{Z}$-bilinear R-balanced $\alpha : M \times N \to G$ there is a unique $\overline{\alpha} : M \otimes_R N \to G$ such that $\alpha = \overline{\alpha} \circ \varphi$.

The tensor product of module homomorphisms. If $f : M_R \to M'_R$ and $g :_R N \to_R N'$ be R-module homorphisms there is a unique $\mathbb{Z}$-bilinear map $f \otimes g : M \otimes_R N \to M' \otimes_R N'$ such that $(f \otimes g)(m \otimes n) = (f(m) \otimes g(n)), \forall m \in M, n \in N$.

In particular, considering R as $R - R$-module, there is a surjective R-module homomorphism, $\mu_R : R \otimes_R N \to RN$ defined by

$$\mu_R \left(\sum_{i=1}^{k} (r_i \otimes n_i) \right) = \sum_{i=1}^{k} r_i n_i.$$

If R is an integral domain a R-module $_R M$ is called

torsion if $\forall m \in M, \exists r \in R^* : rm = 0$ and

divisible if $\forall m \in M, \forall r \in R^*, \exists m_1 \in M : m = rm_1$ (i.e. $rM = M, \forall r \in R^*$).

Ex. 9.1 For $n \in \mathbb{N}^*$ verify $\mathbb{Z} \otimes_{\mathbb{Z}} \mathbb{Z}_n \cong \mathbb{Z}_n$.

Ex. 9.2 Let $n \in \mathbb{N}^*$. Prove that the following equalities hold:
 (a) $\mathbb{Z}_n \otimes_{\mathbb{Z}} \mathbb{Q} = 0$;
 (b) $\mathbb{Z}_n \otimes_{\mathbb{Z}} (\mathbb{Q}/\mathbb{Z}) = 0$;
 (c) $\mathbb{Q} \otimes_{\mathbb{Z}} (\mathbb{Q}/\mathbb{Z}) = 0$. Generalization.

Ex. 9.3 Show that if $u : n\mathbb{Z} \to \mathbb{Z}$ denotes the inclusion then the homomorphism $u \otimes 1_{\mathbb{Z}} : n\mathbb{Z} \otimes_{\mathbb{Z}} \mathbb{Z}_n \to \mathbb{Z} \otimes_{\mathbb{Z}} \mathbb{Z}_n$ is not injective.

Ex. 9.4 Let R be a ring, I an ideal of R and $_RM$ a left R-module. If $I \cdot M = \left\{ \sum_{k=1}^{n} r_k x_k \middle| r_k \in I, x_k \in M, n \in \mathbb{N}^* \right\}$ show that there is a natural isomorphism of R-modules $(R/I) \otimes_R M \cong M/I \cdot M$.
 Application: $R = \mathbb{Z}$.

Ex. 9.5 Let I, J be ideals of the ring R. Point out an isomorphism of R-modules $R/I \otimes_R R/J \cong R/(I + J)$.

Ex. 9.6 Let $n, m \in \mathbb{N}^*$ and $d = \gcd(m, n)$. Prove that the following properties hold:
 (a) $\mathbb{Z}_n \otimes_{\mathbb{Z}} \mathbb{Z}_m \cong \mathbb{Z}_d$.
 (b) $\mathbb{Z}_n \otimes_{\mathbb{Z}} \mathbb{Z}_m = 0$ iff $d = 1$.
 (c) $\mathbb{Z}_n \otimes_{\mathbb{Z}} \mathbb{Z}_n \cong \mathbb{Z}_n$. Generalization.

Ex. 9.7 Determine a group-isomorphism between $\mathbb{Q} \otimes_{\mathbb{Z}} \mathbb{Q}$ and $\mathbb{Q}$.

Ex. 9.8 Let U, V, W be vector spaces over a field K. Find canonical isomorphisms $K \otimes_K U \cong U, U \otimes_K V \cong V \otimes_K U$ and $(U \otimes_K V) \otimes_K W \cong U \otimes_K (V \otimes_K W)$.

Ex. 9.9 Let U^* be the dual of a vector space over a field K. For two finite dimensional vector spaces prove that $(U \otimes_K V)^* \cong U^* \otimes_K V^*$ and $\hom_K(U, V^*) \cong (U \otimes_K V)^*$ hold.

Ex. 9.10 For a field K prove that $K[X, Y] \cong K[X] \otimes_K K[Y]$.

Ex. 9.11 Establish the following isomorphisms:
  (a) $\mathbb{R} \otimes_{\mathbb{Q}} \mathbb{R} \cong \mathbb{R}$ as $\mathbb{Q}$-vector spaces;
  (b) $\mathbb{C} \otimes_{\mathbb{Q}} \mathbb{C} \cong \mathbb{C} \times \mathbb{C}$ as $\mathbb{Q}$-vector spaces;
  (c) $\mathbb{C} \otimes_{\mathbb{R}} \mathbb{C} \cong \mathbb{R}^4$ as $\mathbb{R}$-vector spaces.

Ex. 9.12 Prove that the map $\mu_{\mathbb{C}} : \mathbb{C} \otimes_{\mathbb{R}} \mathbb{C} \to \mathbb{C}$ induced by multiplication, is a ring homomorphism but is not injective.

Ex. 9.13 Let M, M_1 be right R-modules and N, N_1 be left R-modules and let $f : M \to M_1, g : N \to N_1$ be R-module homomorphisms. If f, g are surjective show that $f \otimes g$ is surjective too and that $\ker(f \otimes g)$ is the subgroup of $M \otimes_R N$ generated by the union of the following sets $\{a \otimes y | a \in \ker(f), y \in N\}$ and $\{x \otimes b | x \in M, b \in \ker(g)\}$.

Ex. 9.14 Let R be a ring and let I be an ideal of R such that $I^2 \neq I \neq R$. Show that $i \otimes 1_{R/I}$, where $i : I \to R$ is the inclusion, is not injective

Chapter 10

Artinian and Noetherian Rings

A ring R is called *left (right) artinian* if the set of all the left (right) ideals of R satisfies the *dcc (descending chain condition)* i.e. each strictly descending chain of left (right) ideals is finite (or equivalently, each non-void family of left (right) ideals contains a minimal element). Similarly, a ring is called *left (right) noetherian* if the set of all the left (right) ideals satisfies the *acc (ascending chain condition)*.

Theorem of Hopkins. *Every right artinian ring is right noetherian.*

In a right noetherian ring the prime radical (the intersection of all the prime ideals) is the largest nilpotent right ideal and the largest nil left ideal.

Any left nilideal in a left artinian ring is nilpotent.

Theorem of Hilbert. *If R is right noetherian ring with identity then $R[X]$ is also right noetherian.*

A ring is right noetherian iff each right ideal is finitely generated.

Ex. 10.1 Show that $\mathbb{Z}[i]$ is a noetherian but not an artinian ring.

Ex. 10.2 For two rings R and S and a $R - S$-bimodule M let $T = \begin{pmatrix} R & M \\ 0 & S \end{pmatrix}$ be a ring with componentwise defined addition and a multiplication given by $\begin{pmatrix} r & x \\ 0 & s \end{pmatrix} \cdot \begin{pmatrix} r' & x' \\ 0 & s' \end{pmatrix} = \begin{pmatrix} rr' & rx' + xs" \\ 0 & ss' \end{pmatrix}$. Prove that:

(a) All the left ideals in T have the form $\left\{ \begin{pmatrix} r & m \\ 0 & s \end{pmatrix} \right\}$ where $(r,m) \in N$ a R-submodule of $R \oplus M$, $s \in I$ a left ideal in S, $MI \subseteq N$.

(b) The ring T is left (resp. right) noetherian iff R and S are left (resp.right) noetherian rings and M is a left (resp. right) noetherian R-module (resp. S-module).

Ex. 10.3 Show that $\begin{pmatrix} \mathbb{Z} & \mathbb{Q} \\ 0 & \mathbb{Q} \end{pmatrix}$ is a right noetherian ring but not a left noetherian ring.

Ex. 10.4 Show that $\begin{pmatrix} \mathbb{Q} & \mathbb{R} \\ 0 & \mathbb{R} \end{pmatrix}$ is a right artinian but not a left artinian ring.

Ex. 10.5 Show that $\begin{pmatrix} \mathbb{Q} & \mathbb{Q} \\ 0 & \mathbb{Z} \end{pmatrix}$ is a left noetherian ring but not a right noetherian ring.

Ex. 10.6 Show that $\begin{pmatrix} \mathbb{R} & \mathbb{R} \\ 0 & \mathbb{Q} \end{pmatrix}$ is a left artinian but not a right artinian ring.

Ex. 10.7 Let K be a field. Show that $K[X]$ is a noetherian but not an artinian ring.

Ex. 10.8 Show that if $R[X]$ is a noetherian ring then R has the same property.

Ex. 10.9 Let $\mathbb{Q}\langle X,Y \rangle = \mathbb{Q}[[X,Y]]$ the ring of formal series over $\mathbb{Q}$ and $R = \mathbb{Q}[[X,Y]]/(X^2, XY)$. Show that R is not artinian.

Ex. 10.10 Let R be a ring. Show that R is left artinian (noetherian) iff $\mathcal{M}_n(R)$ is left artinian (noetherian).

Ex. 10.11 Verify that each finite product of artinian (noetherian) rings is artinian (noetherian).

Ex. 10.12 Show that an integral ring (without zero divisors) with identity is left artinian iff it is a division ring.

Ex. 10.13 We say that a ring R satisfies the *annihilator conditions* if $l(r(A)) = A, r(l(B)) = B$ hold for every left ideal A and for every right ideal B (see 2.3). Show that if a left noetherian ring R which satisfies the annihilator conditions then it is right artinian.

Ex. 10.14 If a ring R satisfies the annihilator conditions and any one of the conditions - left noetherian, right noetherian, left artinian or right artinian - then it satisfies all four conditions.

Ex. 10.15 Let R be a commutative ring with identity. If R has ideals which are not finitely generated, prove that the set $\mathcal{NF}$ of all the not finitely generated ideals ordered by set inclusion is inductive (i.e. each chain has an upper bound) and the maximal elements in this ordered set are prime ideals.

Ex. 10.16 Prove that a commutative ring R with identity is noetherian iff each prime ideal I is finitely generated.

Ex. 10.17 Let U, V be ideals in a commutative ring R. If U is finitely generated and $R/U, R/V$ are artinian (noetherian) rings then prove that $R/(U \circ V)$ has the same property.

Ex. 10.18 In an artinian commutative ring R prove that each prime ideal is maximal.

Ex. 10.19 Let R be a commutative ring. Prove that R is artinian iff R is noetherian and each prime ideal is maximal.

Ex. 10.20 In a left artinian ring R show that a left ideal L is nilpotent (i.e. there is a $n \in \mathbb{N}^*$ such that $L^n = (0)$) iff L is nil (that is, each element in it is nilpotent). (Hint: see 17.12).

Ex. 10.21 In a right noetherian ring show that every nil ideal is nilpotent.

Ex. 10.22 In a left artinian ring R show that:
    (a) there are maximal nilpotent left ideals;
    (b) there is a greatest nilpotent left ideal N;
    (c) N is actually two-sided;
    (d) $N = rad(R)$.

Chapter 11

Socle and Radical

The *left (right) socle* of a ring R is the sum of all the minimal (simple) left (right) ideals of R. For commutative rings we shall denote this by $s(R)$.

The *left (right) (Jacobson) radical* of a ring R is the intersection of all the maximal left (right) ideals of R. For commutative rings we shall denote this by $rad(R)$.

An element r in a ring R is called *left (right) quasi-regular* if there is an element $s \in R : r + s - sr = 0$ (resp. $r + s - rs = 0$).

A right (left) ideal I is called

essential (or *large*) if for every right ideal $J \neq 0 \Rightarrow I \cap J \neq 0$;

superfluous (or *small*) if for every right ideal $J, I + J = R \Rightarrow J = R$.

Ex. 11.1 If R is an integral domain then R is a field iff $s(R) \neq 0$.

Ex. 11.2 Compute the socle $s(\mathbb{Z}_n)$.

Ex. 11.3 If $I \subset J \subset K$ show that I is essential in K iff I is essential in J and J is essential in K.

Ex. 11.4 If $f : R \to R'$ is a ring homomorphism and J is essential in R' prove that the preimage $f^{-1}(J)$ is essential in R.

Ex: 11.5 Prove that the intersection of all the essential ideals is the socle of a ring.

Ex. 11.6 Show that if I is superfluous then $I \subseteq rad(R)$.

Ex. 11.7 In a ring R with identity, for a right ideal I, show that if $1 + I$ consists only of units then $I \subseteq rad(R)$.

Ex. 11.8 If I is a finitely generated right ideal of R and $I \subseteq rad(R)$ then I is superfluous.

Ex. 11.9 Prove that the sum of all the superfluous right ideals is the radical of a ring.

Ex. 11.10 Compute the radical $rad(\mathbb{Z}_n)$.

Ex. 11.11 In an arbitrary ring show that if $rad(R) \neq R$ then
$$rad(R) = \{r \in R | \forall x, y \in R : xry \text{ is right quasi-regular } \}.$$

Ex. 11.12 In a ring with identity R show that $rad(R) =$
$$\{r \in R | \forall s \in R \text{ the element } 1 - rs \text{ is right invertible} \}.$$

Ex. 11.13 Verify that $rad(\mathcal{M}_n(R)) = \mathcal{M}_n(rad(R))$.

Ex. 11.14 Show that the radical of a ring contains no nonzero idempotent elements. Then compute $rad(\mathcal{M}_n(K))$ and $rad(B)$ for an arbitrary ring K with identity and for a Boole ring B.

Ex. 11.15 For a field K consider $V = \left\{ \begin{pmatrix} a & 0 & 0 \\ b & a & 0 \\ c & 0 & a \end{pmatrix} \mid a,b,c \in K \right\}$

subring in $\mathcal{M}_3(K)$. Show that V is a commutative ring and compute the radical of V.

Ex. 11.16 Compute the radical of the ring $\hat{\mathbb{Z}}_{(p)}$ of all the p-adic integers.

Ex. 11.17 For a field K and $T = \{(a_{ij}) \in \mathcal{M}_n(K) \mid i < j \Rightarrow a_{ij} = 0\}$ compute $rad(T)$ and show that $T/rad(T)$ is a commutative ring.

Ex. 11.18 Let $f : R \to R'$ be a ring homomorphism. Verify the inclusion $f(rad(R)) \subseteq rad(R')$.

Ex. 11.19 Prove that for a commutative ring with identity R, the radical $rad(R[X])$ and the nilradical $\mathcal{N}(R[X])$ (see 13.11) are equal.

Ex. 11.20 If $r \in R$ and $r \in rad(R[X])$ then prove that r is nilpotent.

Ex. 11.21 If R has characteristic 0 then $rad(R[X]) \neq 0$ implies $R \cap rad(R[X]) \neq 0$ (using the natural identification).

Ex. 11.22 Compute $rad(K\langle X \rangle)$, the ring of formal power series in one indeterminate with coefficients in a division ring K (also denoted by $K[[X]]$).

Chapter 12

Semisimple Rings

A ring R is called *left (right) semisimple* if it is a direct sum of minimal left (right) ideals.

A ring is left (right) semisimple iff it coincides with its corresponding socle.

A ring R is left (right) semisimple iff every left (right) ideal in R is a direct summand.

Ex. 12.1 Show that a product of simple rings need not be semisimple.

Ex. 12.2 Prove that a ring is semisimple iff it has no right essential ideals.

Ex. 12.3 If I is a proper ideal in a semisimple ring R, show that R/I is semisimple too.

Ex. 12.4 Every subring of a semisimple ring has to be semisimple ?

Ex. 12.5 Let $f : R \to R'$ a surjective ring homomorphism of rings with identity. Prove that R' is a semisimple ring iff $_R R'$ (via f) is a semisimple left R-module.

Ex. 12.6 Find necessary and sufficient conditions in order $\mathbb{Z}_n$ to be a semisimple ring..

Ex. 12.7 For a family $\{R_i\}_{i \in I}$ of rings show that $\prod_{i \in I} R_i$ is semisimple iff I is finite and each R_i is semisimple.

Ex. 12.8 Is a union of a chain of semisimple rings semisimple ?

Ex. 12.9 Let A be an ideal in a semisimple ring R. Find a central idempotent $e \in R$ such that $A = eA = Ae$.

Ex. 12.10 Let R be a semisimple ring and $e \in R, e^2 = e \neq 0$. Show that Re is a minimal left ideal of R iff eRe is a division ring.

Ex. 12.11 Prove that a ring with identity is left semisimple iff it is right semisimple.

Ex. 12.12 A ring R is left semisimple iff $rad(R) = 0$ and R is left artinian.

Ex. 12.13 Let R be a ring; show that R is semisimple iff $\mathcal{M}_n(R)$ is semisimple.

Ex. 12.14 Prove that for a ring with identity R the following conditions are equivalent: (a) R is semisimple;
  (b) R is left noetherian and regular von Neumann (see 17.14);
  (c) R is left noetherian and regular von Neumann.

Ex. 12.15 Prove that for a Boole ring R the following conditions are equivalent:
  (a) R is artinian;
  (b) R is noetherian;
  (c) R is semisimple;
  (d) R is finite.

Ex. 12.16 Show that the ring of the endomorphisms of a semisimple module is regular von Neumann (see 17.14).

Ex. 12.17 Let $f : R \to R'$ be a ring homomorphism of rings with identity. If $R/rad(R)$ is semisimple show that $rad(f(R)) = f(rad(R))$.

Ex. 12.18 If R satisfies the descending chain condition on principal right ideals and R is a *semiprime* ring, i.e. has no nonzero nilpotent ideals, then prove that R is semisimple.

Ex. 12.19 Let $R = \left\{ \frac{m}{n} \in \mathbb{Q} \mid (6; n) = 1 \right\}$. (a) Prove that R is a ring and that $2R$ and $3R$ are the only maximal ideals. (b) Show that $R/rad(R)$ is semisimple but idempotents do not lift modulo $rad(R)$.

Chapter 13

Prime Ideals, Local Rings

A proper ideal P of a ring R is called

prime if for each two ideals I, J in R the inclusion $I \circ J \subseteq P$ implies $I \subseteq P$ or $J \subseteq P$;

semiprime if it is an intersection of prime ideals.

In a ring with identity every maximal ideal is prime.

The *prime radical* $\mathcal{R}(R)$ is the intersection of all the prime ideals of R.

A ring R is called *prime (semiprime)* if (0) is a prime (semiprime) ideal in R.

A nonvoid subset S of a ring R with identity is called a *multiplicative system* if $1 \in S, 0 \notin S$ and S is closed under multiplication.

A ring R is called *local* if it has a unique maximal left ideal. *A ring is local iff its set of non-units is closed under addition.*

Ex. 13.1 In $\mathbb{Z}[i]$, the ring of Gauss integers, show that (3) and $(1+i)$ are prime ideals but (2) is not a prime ideal.

Ex. 13.2 Point out an infinite number of prime ideals in $\mathbb{Z}[i]$.

Ex. 13.3 Let K be a division ring, $U_{21}(K) = \{(a_{ij}) \in M_2(K) | a_{21} = 0\}$ and $i : U_{21}(K) \to M_2(K)$ the inclusion homomorphism. Prove that (0) is prime in $M_2(K)$ but $i^{-1}(0)$ is not prime in $U_{21}(K)$.

Ex. 13.4 Show that if an ideal I of R is maximal relative to the property $I \cap S = \emptyset$ then I is a prime ideal.

Ex. 13.5 If I is an ideal of a commutative ring R, show that $\sqrt{I}$ (see also 2.15) is the intersection of all the prime ideals that include I.

Ex. 13.6 Let K_4 denote the field with four elements (see 1.9).
    (a) Compute the characteristic of K_4;
    (b) Show that both elements from K_4 which don't belong to the prime subfield verify the equality $x^2 = x + 1$;
    (c) Show that (2) is a prime (principal) ideal of the integral domain $\mathbb{Z}[\omega] = \{a + b\omega | a, b \in \mathbb{Z}\}$ (here $1 \neq \omega \in \mathbb{C}$);
    (d) Verify the isomorphism $K_4 \cong \mathbb{Z}[\omega] / (2)$.

Ex. 13.7 In the ring $4\mathbb{Z}$ let $A = (8)$ is maximal in $4\mathbb{Z}$ but $4\mathbb{Z}/A$ is no field.

Ex: 13.8 If $A = \left\{ \begin{pmatrix} a & b \\ d & c \end{pmatrix} | a, b, c, d \in p\mathbb{Z}, p \text{ a prime number} \right\}$ prove that A is maximal ideal in $M_2(\mathbb{Z})$ but $M_2(\mathbb{Z})/A$ is no field.

Ex. 13.9 Let R be a commutative ring with identity. Show that an ideal $M \neq R$ is maximal iff for every $x \notin M$ there is an element $r \in R$ such that $1 - rx \in M$.

Ex. 13.10 Give an example of a ring (without identity) such that not every maximal ideal is prime.

Ex. 13.11 Let $\mathcal{R}(R)$ be the intersection of all the prime ideals of R (the prime radical of R). In a commutative ring show that $\mathcal{R}(R)$ is the set $N(R)$ of all the nilpotent elements of R (the largest nil ideal, also called *the nilradical of* R).

Ex. 13.12 Show that $\mathcal{R}(R/\mathcal{R}(R)) = 0$.

Ex. 13.13 Let R be an integral domain. Show that in $R[X]$ the intersection of all the maximal ideals is the zero ideal.

Ex. 13.14 In $\mathbb{C}[X, Y]$ prove that the set of all the polynomials with zero constant term form a maximal ideal.

Ex. 13.15 Which of the following ideals of $\mathbb{Q}[X, Y]$ are prime respectively maximal: $(X^2); (X - 2, Y - 3); (Y - 3); (X^2 + 1); (X^2 - 1); (X^2 + 1, Y - 3)$?

Ex. 13.16 Show that the ideal (n, X) is prime in $\mathbb{Z}[X]$ iff n is prime.

Ex. 13.17 If R is a ring with identity such that for each $x \in R$ there is a $n(x) \in \mathbb{N}^*, n(x) > 1$ with the property $x^{n(x)} = x$ show that in R every prime ideal is also maximal.

Application: a Boole ring.

Ex. 13.18 Let R be commutative ring with identity and $Spec(R)$ the set of all the prime ideals $(\neq R)$ of R. For $X \subseteq R$ we denote by $V(X) = \{P \in Spec(R) | X \subseteq P\}$. Verify the following properties:
  (a) if $A = (X)$ then $V(A) = V(X)$;
  (b) $V(0) = Spec(R), V(1) = \emptyset$;
  (c) $V(\bigcup_{i \in I} X_i) = \bigcap_{i \in I} V(X_i)$ for each family $\{X_i\}_{i \in I}$ of subsets of R;
  (d) $V(A \cap B) = V(A.B) = V(A) \cup V(B)$ for every A, B ideals of R;
  (e) the subsets $\{V(X) | X \subseteq R\}$ define a topology (called *Zariski*) in terms of the closed subsets on $Spec(R)$ (also called *the prime spectrum of* R).

Ex. 13.19 Determine $Spec(\mathbb{Z})$ and $Spec(\mathbb{R})$.

Ex. 13.20 For $r \in R$ denote by $U_r = Spec(R) \setminus V(\{r\})$. Show that $\{U_r | r \in R\}$ form a base of open sets in the Zariski topology of R (see the above exercises). Moreover, verify:

(a) $U_r \cap U_s = U_{r.s}$ holds for each $r, s \in R$;

(b) $U_r = \emptyset$ iff r is nilpotent in R;

(c) $U_r = Spec(R)$ iff r is a unit;

(d) $U_r = U_s$ iff $\sqrt{(r)} = \sqrt{(s)}$ for each $r, s \in R$;

(e) $Spec(R)$ and the subsets U_r $(r \in R)$ are (quasi)compact (not necessarily Hausdorff).

Ex. 13.21 Let $f : R \to R'$ be a ring homomorphism and $P' \in Y = Spec(R')$. Prove that $f^{-1}(P') \in X = Spec(R)$ and hence define $f^* : Spec(R') \to Spec(R)$ in the following way $f^*(P') = f^{-1}(P')$. Verify the following properties:

(a) if $r \in R$ then $(f^*)^{-1}(U_r) = U'_{f(r)} = Spec(R') \setminus V(\{f(r)\})$ that is, f^* is continuous in the corresponding Zariski topologies;

(b) if A is an ideal in R then $(f^*)^{-1}(V(A)) = V(f(A).R')$;

(c) if A' is an ideal in R' then $\overline{f^*(V(A'))} = V(f^{-1}(A'))$;

(d) if f is surjective, f^* is a homeomorphism from $Spec(R')$ onto $V(\ker f) \subseteq Spec(R)$;

(e) if f is injective then $f^*(Spec(R'))$ is dense in $Spec(R)$.

Ex. 13.22 Let R be a Boole ring (see 1.23). Prove that:

(a) for each $r \in R$, U_r is open and closed in $Spec(R)$;

(b) for each finite subset $F \subseteq R$ there is an element $r \in R$ such that $\bigcup_{x \in F} U_x = U_r$;

(c) each open and closed subset from $Spec(R)$ has the form U_r for a suitable $r \in R$;

(d) $Spec(R)$ is a compact (i.e. quasicompact and Hausdorff) topological space..

Ex. 13.23 Show that a ring is local iff the set of all the non-units forms an ideal in R.

Ex. 13.24 Show that $\mathbb{Q}^{(p)}$ (see 2.27) is a local ring.

Ex. 13.25 Let R a commutative ring with identity and M a maximal ideal of R such that $a + 1$ is a unit for every $a \in M$. Show that R is local.

Ex. 13.26 Let R be a commutative ring with identity and M a maximal ideal of R. Prove that for each $n \in \mathbb{N}^*$ the ring $R/(M^n)$ is local.

Chapter 14

Polynomial Rings

For an arbitrary ring with identity R, we denote by $R[X]$ the *ring of polynomials of indeterminate X over R*.

If R' is a ring with identity and R is a subring of R' which includes the identity, for each $c \in R'$ such that $cr = rc, \forall r \in R$ there exists a unique unital ring homomorphism $E_c : R[X] \to R'$ such that $E_c(r) = r, \forall r \in R$ and $E_c(X) = c$.

Universality theorem. *If R and R' are rings with identity then for each unital ring homomorphism $f : R \to R'$ and each element $c \in R'$ such that $c.f(r) = f(r).c, \forall r \in R$ there is a unique $f' : R[X] \to R'$ such that $f'(r) = f(r), \forall r \in R$ and $f'(X) = c$.*

If R is a unique factorization domain, a polynomial $f \in R[X]$ is called *primitive* if no prime element divides all its coefficients.

Ex. 14.1 Give an example of a polynomial of degree 2 with 4 zeros over a commutative ring (which is not a field).

Ex. 14.2 Give an example of a polynomial of degree 2 with an infinite number of zeros over a non-commutative (infinite) division ring.

Ex. 14.3 Characterize the fields such that the classical formula which gives the zeros of a degree 2 polynomial holds.

Ex. 14.4 Consider $f = X^4 - 4, g = X^2 + 2$ in $\mathbb{Q}[X]$. Verify that the map $F : \mathbb{Q}[X]/(f) \rightarrow \mathbb{Q}[X]/(g), F(h + (f)) = h + (g)$ is a well-defined ring homomorphism. Prove that $\ker(F)$ is a maximal ideal in $\mathbb{Q}[X][X]/(f)$.

Ex. 14.5 If K is a field show that (X) is a maximal ideal in $K[X]$.

Ex. 14.6 Let D be an integral domain and a an irreducible (in the divisibility sense) element. Prove that (a, X) in $D[X]$ is not principal.

Ex. 14.7 Is the polynomial $X^3 + X^2 + X + \bar{1}$ divisible by $X^2 + \bar{3}X + \bar{2}$ as elements in the polynomial ring over either $\mathbb{Z}_3$, $\mathbb{Z}_5$ or $\mathbb{Z}_7$?

Ex. 14.8 Study the reducibility of the polynomials $X^2 + \bar{1}$ and $X^3 + X + \bar{2}$ in $\mathbb{Z}_3[X]$ respectively $\mathbb{Z}_5[X]$.

Ex. 14.9 Does an element $a \in \mathbb{Z}_5$ exist, such that $X^4 + aX + \bar{1}$ is irreducible in $\mathbb{Z}_5[X]$?

Ex. 14.10 Verify that the polynomials $f = X^5 + X^3 + X$ and $g = X^5 + \bar{2}X \in \mathbb{Z}_3[X]$ induce the same polynomial function.

Ex. 14.11 Let R be a commutative ring with identity and I an ideal in R. Verify that $I_* = \{a_0 + a_1X + .. + a_nX^n \in R[X] | a_0 \in I\}$ is an ideal in $R[X]$. Moreover, prove that I is prime (or maximal) iff I_* is prime (resp. maximal).

Ex. 14.12 Prove that the ideal (n, X) generated by $n \in \mathbb{Z}$ and X is principal in $\mathbb{Z}[X]$ iff $n \in \{-1, 0, 1\}$.

Ex. 14.13 In $\mathbb{Z}[X]$ show that the set of all the polynomials with the constant term even number is an ideal which is not principal.

Ex. 14.14 Prove that in any ring of polynomials $R[X]$ the subset of all the polynomials with zero degree 1 coefficient is a subring. What about the case of zero degree 2 coefficient ?

Ex. 14.15 Let R be a commutative ring with identity and $R[X]$ the corresponding ring of polynomials. Describe the units and the nilpotent elements in $R[X]$.

Ex. 14.16 Let R be a commutative ring with identity. Prove that the following two conditions are equivalent:
  (a) 0 is the unique nilpotent element of R;
  (b) every unit in $R[X]$ has zero degree.

Ex. 14.17 Is the quotient ring $\mathbb{C}[X] / (X^2 + 1)$ an integral domain ? (Here $(X^2 + 1)$ denotes the principal ideal generated by X^2+1 in $\mathbb{C}[X]$).

Ex. 14.18 Verify the following ring isomorphism $\mathbb{C} \cong \mathbb{R}[X] / (X^2+1)$.

Ex. 14.19 Verify the following ring isomorphism $\mathbb{Z}[X] / (n, X) \cong \mathbb{Z}_n$.

Chapter 15

Rings of Quotients

In this chapter we deal only with commutative rings with identity.

A nonvoid subset $S \subseteq R$ is called a *multiplicative system* if $S \cdot S \subseteq S$ and $1 \in S$.

A multiplicative system S of a ring R is called *saturated* if $ss' \in S \Leftrightarrow s, s' \in S$.

For a multiplicative system S in a ring R we define on $R \times S$ the following equivalence relation: $(x, s) \sim (y, t) \Leftrightarrow \exists u \in S : u(tx - sy) = 0$. The corresponding quotient structure $R \times S / \sim$ denoted R_S or $S^{-1}R = \left\{ \frac{x}{s} \overset{not}{=} \overline{(x,s)} \,|\, x \in R, s \in S \right\}$ has a natural structure of commutative ring with identity called the *ring of quotients* of R relative to (or with denominators in) S given by $\frac{x}{s} + \frac{y}{t} = \frac{tx+sy}{st}$, $\frac{x}{s} \cdot \frac{y}{t} = \frac{xy}{st}$.

The map $\varphi : R \to R_S, \varphi(x) = \frac{x}{1}$ is a unital ring homomorphism, injective iff $S \subseteq \{$ non zero-divisors of $R\}$ and $\varphi(S) \subseteq U(S^{-1}(R))$.

The universality theorem. *For every ring R' and every ring homomorphism $f : R \to R'$ such that $f(S) \subseteq U(R')$ there is a unique ring homomorphism $\overline{f} : R_S \to R'$ such that $f = \overline{f} \circ \varphi$.*

Consequently, (R_S, φ) is unique up to a ring homomorphism.

If $S = \{$non zero-divisors$\}$ we use the notation $Q(R)$, called the *classical (total) ring of quotients*. For an integral domain (here $S = R^*$) $Q(R)$ is the *field of fractions*; for a prime ideal P if $S = R \backslash P$ we obtain the *localization* of R in P, denoted R_P a *local ring*.

Ex. 15.1 Prove that each subring with identity of $\mathbb{Q}$ has the form $\mathbb{Z}_S$ for a suitable multiplicative system $S \subset \mathbb{Z}$.

Ex. 15.2 Determine the ring of quotients of the ring $\mathbb{Z}[i]$ of the Gauss integers.

Ex. 15.3 Let p be a prime number and $\mathbb{Q}^{(p)} = \left\{ \frac{m}{n} \in \mathbb{Q} | (n; p) = 1 \right\}$ (see 2.27). Show that $\mathbb{Q}^{(p)}$ is an integral domain and determine its ring of quotients.

Ex. 15.4 If K is a field and R a subring of K then the field of fractions $Q(R)$ is ring isomorphic with the subfield of K generated by R (i.e. the smallest field that contains R as subring).

Ex. 15.5 If R is a principal ring and $0 \notin S$ is a multiplicative system in R prove that R_S is principal too.

Ex. 15.6 Show that S is saturated iff $R \backslash S$ is a union of prime ideals.

Ex. 15.7 Let $a \in R$ be a non-zero divisor in a commutative ring with identity. If $S = \{1, a, a^2, .., a^n, ..\}$ verify $R_a = R_S \cong R[X]/(aX - 1)$.

Ex. 15.8 For the canonical homomorphism $f : R \to R_S$ show that the right cancellability property holds (f is an *epimorphism* in the category of of rings).

Ex. 15.9 With the above notations, if $q_1, q_2, .., q_n \in R_S$ then there is an element $s \in S$ such that $q_1 s, q_2 s, .., q_n s \in f(R)$.

Ex. 15.10 Prove that the canonical morphism $f : R \to R_S$ is an isomorphism iff $S \subseteq U(R)$, the units of R. In particular, the field of quotients of a field is isomorphic with itself.

Ex. 15.11 If D is an integral domain and S a multiplicative system in D then the canonical homomorphism $f : D \to D_S$ is an isomorphism iff for each element $q \in D_S$ there is an monic polynomial $p_q \in D[X]$ such that $p_q(q) = 0$.

Ex. 15.12 If R is a right noetherian ring verify that R_S is also right noetherian.

Ex. 15.13 For $S = 2\mathbb{Z}^* \cup \{1\}$ consider the ring of quotients $\mathbb{Z}_S = S^{-1}\mathbb{Z}$. Show that $\mathbb{Z}_S \cong \mathbb{Q}$.

Ex. 15.14 Show that the canonical homomorphism $\varphi : R \to R_S$ is surjective iff R_S is cyclic as a R-module (via φ).

Ex. 15.15 Let $n \in \mathbb{N}, n \geq 2$ and S be a multiplicative system in $\mathbb{Z}_n$. If $f : \mathbb{Z}_n \to (\mathbb{Z}_n)_S$ is the canonical homomorphism show that $(\mathbb{Z}_n)_S \cong \mathbb{Z}_n / \ker f$.

Ex. 15.16 If $S = \mathbb{Z}^*$ show that $f : \left(\prod_{\mathbb{N}} \mathbb{Z}\right)_S \to \prod_{\mathbb{N}} (\mathbb{Z})_S$, the canonical homomorphism, is not surjective.

Ex. 15.17 Let I be an ideal in a ring R with identity. Show that $S = 1 + I$ is a multiplicative system and I_S is contained in $Rad(R_S)$.

Ex. 15.18 Each injective ring homomorphism $D \to D'$ of integral domains induces an (injective) ring homomorphism $Q(D) \to Q(D')$ of division rings.

Ex. 15.19 If $D \cong D'$ are integral domains then the fields of quotients $Q(D) \cong Q(D')$.

Ex. 15.20 (a) If P is a prime ideal in a commutative ring with identity R, show that $R \backslash P$ is a multiplicative system.

(b) For simplification R_P denotes the ring of quotients $R_{R \backslash P}$. Show that R_P is a local ring. How can its quotient ring modulo its maximal ideal be described ?

Ex. 15.21 With the above notations show that for all the prime ideals P of a ring R, R_P has no non-zero nilpotent elements iff R has no non-zero nilpotent elements.

Ex. 15.22 With the above notations, if R_P are integral domains, has R the same property ?

Chapter 16

Rings of Continuous Functions

Let X be a nonvoid set. A *topology* on X is a family $\tau \subset \mathcal{P}(X)$ of subsets which is closed under finite intersections and arbitrary unions. The pair (X, τ) is called a *topological space*, the elements of τ are called the *open* sets of the space. If τ_1 and τ_2 are topologies on X and $\tau_1 \subset \tau_2$ we say that τ_2 is *finer* than τ_1 or that τ_1 is *coarser* than τ_2. It exists in X a finest topology $\tau^\circ = \mathcal{P}(X)$ (called the *discrete* topology) and a coarsest one $\tau_\circ = \{\emptyset, X\}$ (called the *indiscrete* topology). The elements of $\mathcal{F} = \{X \setminus G | G \in \tau\}$ are called the *closed* sets.

A topological space is called *disconnected*, if it exists a nonvoid open set with an open, nonvoid complement; otherwise it is called *connected*. The space is called *countably compact* if every countable open cover of X has a finite subcover.

We denote by $\mathbb{R}^X$ the set of all real-valued functions on the set X and with $C(X)$ $(C^*(X))$ the set of all the continuous (bounded) real-valued functions on a topological space X. This function sets are provided with the usual operations of addition and multiplication. For $\alpha \in \mathbb{R}$ we denote with $\underline{\alpha}$ the constant function with value α.

We recall the following property which can be used in order to prove the continuity of a map: *if $f \in \mathbb{R}^X$ and $\{G_i : i \in I\}$ is an open cover of the space X and each restriction $f \mid G_i$ is continuous, then f is continuous on X.* The inequality $f \leq g$ means $f(x) \leq g(x)$ for each $x \in X$.

$(\mathbb{R}^X, \leq)$, $(C(X), \leq)$ and $(C^*(X), \leq)$ are lattices (and notice that $f \wedge g = \min(f, g)$ and $f \vee g = \max(f, g)$).

If $C(X) = C^*(X)$, then X is called *pseudocompact*.

We associate to each $f \in C(X)$ his *zero-set*: $Z(f) = f^{-1}(0)$, the preimage. Each zero-set is obviously closed in X. The set of all the zero-sets of the space X will be denoted by $Z(X)$.

A ring R is the *direct sum* of two subrings S_1, S_2, if $S_1 \cap S_2 = \{0\}$, and for each $a \in R$ it exists $a_1 \in S_1$ and $a_2 \in S_2$ with $a = a_1 + a_2$. We denote this by $R = S_1 \oplus S_2$.

In the following exercises, the main concern is the relations between the topological properties of X and the algebraic properties of $C(X)$ and $C^*(X)$.

Ex. 16.1 Let R be a ring and X a nonvoid set. Denote by $R^X = \{f | f : X \to R\}$ the set of all the maps together with the usual operations of addition and multiplication. Prove that:

(a) $\left(R^X, +, \cdot\right)$ is a ring;

(b) R commutative $\Leftrightarrow R^X$ commutative;

(c) R has identity $\Leftrightarrow R^X$ has identity;

(d) If R has identity then $f \in R^X$ is a unit iff $f(x)$ is a unit of R for each $x \in X$;

(e) If R and X have both at least two elements then R^X has zero divisors;

(f) If R is a Boole ring then R^X is a Boole ring too.

Ex. 16.2 The set $C(X)$ forms a subring with identity and a sublattice of $\mathbb{R}^X$. The set $C^*(X)$ forms a subring with identity and a sublattice of $C(X)$.

Ex. 16.3 (a) For each topological space X the following inclusions $C_o(X) \subset C^*(X) \subset C(X) \subset C^\circ(X)$ hold, where $C_o(X) = \{\underline{\alpha} | \alpha \in \mathbb{R}\}$ and $C^\circ(X) = \mathbb{R}^X$.

(b) If X is an indiscrete space (i.e $\tau = \tau_o$), then $C(X) = C_o(X)$. The converse is not true. [Find a topological space (X, τ) with $\tau \neq \tau_o$ such that $C(X) = C_o(X)$ and X is: (i) finite, (ii) infinite.]

(c) $C(X) = C^\circ(X)$ if and only if X is a discrete space (i.e.$\tau = \tau^\circ$).

Ex. 16.4 Describe $C(X)$ for each possible topology in the set

a) $X = \{a, b\}$;

b) $X = \{a, b, c\}$.

Ex. 16.5 If $f \in C(X)$, then also $|f| \in C(X)$. Find an algebraic way

a) to check for $f \in C(X)$ whether $f \geq 0$;

b) to characterize for a given $f \in C(X)$ the element $|f|$.

Ex. 16.6 Verify that the following conditions are equivalent:

(i) X is disconnected,

(ii) in $C(X)$ the identity has at least three square roots,

(iii) there exist at least three idempotents in $C(X)$,

(iv) $C(X)$ is the direct sum of two subspaces.

Ex. 16.7 If X is countable compact, show that X is pseudocompact.

Ex: 16.8 The converse of 16.7 is false. [Hint: see exercise 16.3(b)]

Ex. 16.9 Let X be a topological space, $f \in C(X)$ and $f^* = (-\underline{1} \vee f) \wedge \underline{1}$. Show that it exists a unit $u > 0$ of $C(X)$ satisfying $f^* = uf$.

Ex. 16.10 Prove that $C(X)$ and $C^*(X)$ have the same zero-sets.

Ex. 16.11 $Z(X)$ is a sublattice of the lattice of all the closed sets of X.

Ex. 16.12 Find a topological space X such that not all the closed sets are zero-sets. [Hint: see exercise 16.3(b)].

Ex. 16.13 Prove that in a metric space each closed space is a zero-set.

Ex. 16.14 Verify or analyse the following assertions:
    (a) f is a unit in $C(X)$ iff $Z(f) = \emptyset$.
    (b) Under which conditions if f a unit in $C^*(X)$?
    (c) Find a space X and $f \in C^*(X)$ with $Z(f) = \emptyset$ such that f is no unit in $C^*(X)$.

Ex. 16.15 Let $f, g \in C(X)$ and let $Z(f)$ be a neighbourhood of $Z(g)$, i.e. there is an open subset G of X such that $Z(g) \subseteq G \subseteq Z(f)$. Show that g divides f in $C(X)$ (i.e. there is a map $h \in C(X)$ such that $f = g \cdot h$).

Ex. 16.16 For a map $f \in C(X)$ (resp. $C^*(X)$) we say that f has the property $\|$ if there is a unit $u \in C(X)$ (resp. $C^*(X)$) such that $|f| = u \cdot f$. Show that:
    (a) Each $f \in C(X)$ has the property $\|$ iff each $f \in C^*(X)$ has the property $\|$;
    (b) If X has the discrete topology, find the elements from $C(X)$ which have the property $\|$;
    (c) Determine the elements of $C(\mathbb{R})$ ($\mathbb{R}$ with the usual topology) which have the property $\|$;
    (d) Find an element in $C(\mathbb{Q})$ ($\mathbb{Q}$ with the usual induced topology) which has not the property $\|$;

(e) If $f \in C(\mathbb{Q})$ and $Z(f)$ is an open subset then f has the property $\parallel$. In the previous conditions find an example of function such that $f \not\leq 0$ and $f \not\geq 0$.

Ex. 16.17 Let X be a topological space and I an ideal in $C(X)$. If we denote by $\mathcal{Z} = Z[I] = \{Z(f)|f \in I\}$, verify the following properties:

[z_1] $\mathcal{Z} \subseteq Z(X)$;

[z_2] $Z_1, Z_2 \in \mathcal{Z} \Rightarrow Z_1 \cap Z_2 \in \mathcal{Z}$;

[z_3] $Z \in \mathcal{Z}, Z \subseteq Z' \Rightarrow Z' \in \mathcal{Z}$.

Ex. 16.18 A family $\mathcal{Z} \subseteq \mathcal{P}(X)$ which satisfies [z_1] $-$ [z_3] (see the previous exercise) is called a *z-filter*. Show that if I is a proper ideal in $C(X)$ then the corresponding z-filter is also *proper*, i.e. $Z[I] \neq \mathcal{Z}$.

Ex. 16.19 If $\mathcal{Z}$ is a z-filter show that $Z^{\leftarrow}(\mathcal{Z}) = \{f \in C(X)|z(f) \in \mathcal{Z}\}$ is an ideal in $C(X)$, proper if $\mathcal{Z}$ is proper.

Ex. 16.20 (a) If M is a maximal ideal in $C(X)$ then $\mathcal{Z} = Z(M)$ is a *maximal* z-filter (i.e. a proper z-filter not properly included in any other proper z-filter).

(b) Conversely, if $\mathcal{Z}$ is a maximal z-filter then $Z^{\leftarrow}(\mathcal{Z})$ is a maximal ideal.

Ex. 16.21 (a) If P is a prime ideal in $C(X)$ then $\mathcal{Z} = Z(P)$ is a *prime z-filter*, i.e. a proper z-filter with the property [z_P] : $Z_1, Z_2 \in Z(P), Z_1 \cup Z_2 \in \mathcal{Z} \Rightarrow Z_1 \in \mathcal{Z}$ or $Z_2 \in \mathcal{Z}$.

(b) If $\mathcal{Z}$ is a prime z-filter then $Z^{\leftarrow}(\mathcal{Z})$ is a prime ideal in $C(X)$.

Ex. 16.22 Prove that every maximal z-filter is prime. Moreover, each maximal ideal in $C(X)$ is prime.

Ex. 16.23 Consider $1_{\mathbb{R}} \in C(\mathbb{R})$ and the principal ideals $I_1 = \left(1_{\mathbb{R}}\right)$ and $I_2 = \left(1_{\mathbb{R}} \cdot 1_{\mathbb{R}}\right)$.

(a) Show that $I_2 \subset I_1, I_2 \neq I_1$ but $Z(I_1) = Z(I_2)$.

Remark. The map Z which associates a z-filter to each ideal is therefore not injective. $Z^{\leftarrow}(\mathcal{Z})$ is the greatest ideal who maps on $\mathcal{Z}$.

(b) Neither I_1 nor I_2 are prime ideals.

(c) Find a maximal ideal which includes I_1 and I_2.

(d) Verify that $Z(I_1)$ is a maximal z-filter.

(e) Determine $Z^{\leftarrow}(Z(I_1))$.

Ex. 16.24 (a) Show that the following subsets are z-filters in $\mathbb{R}$:

$Z_+ = \{\Gamma \in \mathbb{R}| \Gamma \in \mathcal{F}, \exists \varepsilon > 0 : [a, a + \varepsilon[\subset \Gamma\}$ and

$Z_- = \{\Gamma \in \mathbb{R}| \Gamma \in \mathcal{F}, \exists \varepsilon > 0 :]a - \varepsilon, a] \subset \Gamma\}$. Are these prime ?

(b) Show that $Z_0 = \{\Gamma_1 \cap \Gamma_2| \Gamma_1 \in Z_+, \Gamma_2 \in Z_-\}$ is also a z-filter. Is it prime ?

Chapter 17

Special Problems

An idempotent element $e \neq 0$ is called *primitive* if it cannot be written as the sum of two orthogonal idempotent elements.

A ring R is called *regular (von Neumann)* if for every $a \in R$ there is an element $b \in R$ such that $a = aba$ and *strongly regular* if there is a unique element $b \in R$ with this property.

Ex. 17.1 Let R be a ring with identity. Show that if $a, b, a - b$ are units then $a^{-1} - b^{-1}$ is a unit too if: (a) is commutative; (b) is not necessarily commutative.

Ex. 17.2 Let R be a ring with identity such that $r^3 = r, \forall r \in R$. Show that R is commutative.

Ex. 17.3 If $x^6 = x$ holds for every element x in a ring R then show that also $x^2 = x$.

Ex. 17.4 Let R be a ring without nonzero nilpotent elements. If for $a_1, a_2, .., a_n \in R$ the relation $a_1 a_2 .. a_n = 0$ holds prove the following:
 (a) $\forall i \in \{1, 2, .., n\} : a_i a_{i+1} .. a_n a_1 .. a_{i-1} = 0$;
 (b) for every $k \in \mathbb{N}^*, k \leq n - 1$ and for every $b_1, b_2, .., b_k \in R :$ $a_1 b_1 a_2 b_2 .. a_k b_k a_{k+1} .. a_n = 0$;
 (c) for every permutation $\sigma \in S_n : a_{\sigma(1)} a_{\sigma(2)} .. a_{\sigma(n)} = 0$.

Ex. 17.5 Let R be a ring with identity and $n \in \mathbb{N}$ such that $(xy)^k = x^k y^k$ holds for each $x, y \in R$ and $k \in \{n, n+1, n+2\}$. Show that R is commutative.

Ex. 17.6 Show that if a ring R has no nilpotent ideals and for every $r \in R$ there is a number $n(r) \in \mathbb{N}^*$ such that $r^{n(r)} \in Z(R)$ then R is commutative.

Ex. 17.7 Let K be a field. Show that there are no nonzero elements a, b in K such that $2a = 3b = 0$.

Ex. 17.8 Let R be a finite commutative ring with identity and $0 \neq 1$. Compute the product of all the nonzero idempotent elements of R.

Ex. 17.9 Show that if I is a minimal right ideal of a ring R then $I^2 = 0$ or $I = eR$ for a suitable idempotent element $e \in I$.

Ex. 17.10 Verify that $e \neq 0$ is a primitive idempotent iff eRe contains no idempotents other than 0 and e.

Ex. 17.11 If e is a primitive idempotent in a regular (von Neumann) ring, prove that eRe is a division ring.

Ex. 17.12 Let R be a left artinian ring and N a nonnilpotent left ideal. Prove that there is an element $a \in N$ such that $a^n \neq 0$ for every $n \in \mathbb{N}^*$.

Ex. 17.13 If K is a finite field show that each map $f : K \to K$ is polynomial.

Ex. 17.14 Show that for a ring with identity the following conditions are equivalent:
   (i) R is von Neumann;
   (ii) each principal left (right) ideal is a direct summand;
   (iii) each finitely generated left (right) ideal is a direct summand.

Ex. 17.15 In a strongly regular ring show:
   (a) R has no zero divisors;
   (b) $bab = b$ (with the above notations);
   (c) R has identity;
   (d) R is a division ring.

Ex. 17.16 In a regular ring R show that for each element $a \in R$ there is an element $c \in R$ such that $a = aca$ and $c = cac$. Application: the center of a regular ring is also regular.

Ex. 17.17 Show that if a ring with identity is a sum of ideals, then it is a finite sum.

Ex. 17.18 Let H be a nonzero ideal of $\mathbb{Z}[i]$. Prove that the ring $\mathbb{Z}[i]/H$ is finite.

Ex. 17.19 Let R be a commutative ring with identity. Show that the following conditions are equivalent: (i) $Spec(R)$ is not connected;
   (ii) $R \cong U \times V$ for two suitable nonzero rings;
   (iii) There is an idempotent element $e \in R, e \notin \{0, 1\}$.
   Application: $R = \mathbb{Z}_{p^k}[X]/(X^2 - X + p^i)$ for p prime, $i, k \in \mathbb{N}$ and $1 \leq i \leq k$.

Ex. 17.20 (Chinese remainder theorem) Let $I_1, I_2, .., I_n$ be a finite set of ideals in a ring R. Show that the following assertions are equivalent:

(a) The ideals are comaximal in pairs, that is $I_i + I_j = R$ for each $i \neq j$.

(b) For each set of elements $a_1, a_2, .., a_n$ from R show that there is an element $r \in R$ such that $r - a_i \in I_i$ for each $1 \leq i \leq n$.

(c) The canonical map

$$u : R/\bigcap_{i=1}^{n} I_i \to \prod_{i=1}^{n} R/I_i, u\left(a + \bigcap_{i=1}^{n} I_i\right) = (a + I_1, .., a + I_n) \text{ is an}$$

isomorphism.

Ex. 17.21 (Eisenstein criterion) Let R be a unique factorization domain and $f = a_0 + a_1 X + .. + a_n X^n \in R[X], a_n \neq 0$. If p is an irreducible element in R such that p divides $a_0, a_1, .., a_{n-1}$ but p divides not a_n and p^2 divides not a_0 then f is irreducible in $K[X]$, where K is the field of quotients of R. If no prime element divides all the coefficients of f (a *primitive* polynomial) in the above hypothesis f is irreducible also in $R[X]$.

Part II

SOLUTIONS

Chapter 1

Fundamentals

Ex. 1.1 One has only to compute in two different ways (using the distributivity laws) the element: $(1+1)(a+b) = 1(a+b) + 1(a+b) = a+b+a+b$ and $(1+1)(a+b) = (1+1)a + (1+1)b = a+a+b+b$. Now $a+b = b+a$ follows by additive cancellation.

Ex. 1.2 (a) For each $a, b, c \in R$ one has $(a \circ b) \circ c = (a+b-ab) \circ c = a+b+c-ab-ac-bc = a \circ (b+c-bc) = a \circ (b \circ c)$. As for the identity, in this semigroup we have $a \circ 0 = 0 \circ a = a$.

(b) If $QR(R)$ denotes the set of all the quasi-regular elements it is readily checked that $QR(R)$ is closed for the circle composition and contains 0 (indeed, $r \circ s = r' \circ s' = 0 \Rightarrow (r \circ r') \circ (s' \circ s) = 0$ and $0 \circ 0 = 0$). For every quasi-regular element $r \in R$ if $r \circ s = 0$, and $t \circ r = 0$ then $t = t \circ 0 = t \circ (r \circ s) = (t \circ r) \circ s = 0 \circ s = s$ and this is the inverse of r in $QR(R)$.

Ex. 1.3 $(a+b)^2 = (a+b)(a+b) = a^2 + ab + ba + b^2$ so that the stated condition is equivalent (by addition cancellation) with $ab = ba$.

Ex. 1.4 If $(n; m) = 1$ there are integers $s, t \in \mathbb{Z}$ such that $sn + tm = 1$. But then $a = a^1 = a^{sn+tm} = (a^n)^s . (a^m)^t = (b^n)^s . (b^m)^t = b^{sn+tm} = b^1 = b$.

Ex. 1.5 The ring verifications are straightforward. In the first ring $(0,1)^2 = (0,0)$ so that $(0,1)$ is a (twosided) zero divisor. The map $f : (\mathbb{R}^2, +, *) \to (\mathbb{C}, +, \cdot)$ defined by $f(x,y) = x + iy$ is easily checked

to be a ring isomorphism. The above two rings are not isomorphic.
More generally, if R and R' are rings such that R has and R' has
not zero divisors then these two rings are not isomorphic. Indeed, if
$f : R \to R'$ would be an isomorphism (but injective ring homomorphism
= monomorphism suffices) and $ab = 0, a, b \in R$ with $a \neq 0 \neq b$ then
$f(a) \cdot f(b) = 0$ and $f(a) \neq 0 \neq f(b)$, contradiction.

Ex. 1.6 $\emptyset$ and M are the zero respectivelly the identity element, each
non-void disjoint subsets of M are zero divisors and clearly $A \cap A = A$
so that each element is idempotent. All the ring verifications are simple
with only one exception: the associativity of the addition. Let A, B, C
be subsets of M. One has to use the following immediate formula $A +
B = (A \cup B) \cap C_M(A \cap B)$ and to obtain by computation $(A+B)+C =
\{[(A \cup B) \cap C_M(A \cap B)] \cup C\} \cap C_M[(A \cup B) \cap C_M(A \cap B) \cap C] = (A \cup B \cup
C) \cap (A \cup C_M(B) \cup C_M(C)) \cap (C_M(A) \cup B \cup C_M(C)) \cap (C_M(A) \cup C_M(B) \cup C)$
which equals with $A + (B + C)$ by commutativity (which is obvious)
and symmetry.

Ex. 1.7 By the distributivity laws, in order to define a multiplication
$*$ on $\mathbb{Z}$, we remark that it suffices to define $1 * 1$. Indeed, $n * m =
(1 + .. + 1) * (1 + .. + 1) = 1 * 1 + .. + 1 * 1 = nm(1 * 1)$. One easily
checks that for every $k \in \mathbb{Z}$ the multiplication $n * m = nmk$ defines a
ring structure on $\mathbb{Z}$. The only structures with identity are obtained for
$k = 1$ or $k = -1$.

Ex. 1.8 Let $(A, +, \odot)$ a ring with identity $e \in A$ and $(A, +) = (\mathbb{Z}_n, +)$.
$(A, +)$ being an abelian group let $m = ord(e)$. From the well-known
group-theoretic theorem of Lagrange we derive that m divides n. For
each $a \in A$ the following holds $ma = m(e \odot a) = (e \odot a) + .. + (e \odot
a) = (e + .. + e) \odot a = (me) \odot a = \bar{0} \odot a = \bar{0}$ so that for $a = \bar{1}$ we
obtain $m\bar{1} = \bar{m} = \bar{0}$, that is, n divides m. If now $m = n$ we have
$A = \{\bar{0}, e, 2e, .., (n-1)e\}$ and one has only to check that $f : A \to \mathbb{Z}_n$,
defined by $f(ke) = \bar{k}$ for every $ke \in A$, is a ring isomorphism.

Ex. 1.9 In the previous exercise we have seen that on a cyclic group
with four elements, up to an isomorphism, one has only the usual ring
structure on $(\mathbb{Z}_4, +, .)$ But there are (up to a group isomorphism) only

two (abelian) groups with four elements: the cyclic one and the Klein group (or $(\mathbb{Z}_2 \times \mathbb{Z}_2, +)$)i.e. $\mathcal{K} = \{0, a, b, c\}$ with the addition given by table 1.1

Table 1.1:

	0	a	b	c
0	0	a	b	c
a	a	0	c	b
b	b	c	0	a
c	c	b	a	0

The elements a, b, c acting symmetrically, suppose that a is the identity of the four elements ring. All the possible multiplications will be defined by tables like table 1.2

Table 1.2:

	0	a	b	c
0	0	0	0	0
a	0	a	b	c
b	b	y		
c	c			

Denoting $y = b.b$ we have $b.c = b(a + b) = b + y$, $c.b = (a + b)b = b + y$ and $c.c = (a + b)(a + b) = a + y$. Hence there are four possible multiplications on $\mathcal{K}$:

Case 1: $b.b = 0$. We obtain a first ring structure given by the multiplication in table 1.3

Denote this ring by I_4.

Case 2: $b.b = a$. We obtain a ring isomorphic with the above one . Indeed, the function $f : \mathcal{K} \to \mathcal{K}$ defined by $f(0) = 0$, $f(a) = a$, $f(b) = c$, $f(c) = b$ is easily checked to be a ring isomorphism.

Case 3: $b.b = b$. We obtain another ring structure given by the multiplication in table 1.4

Table 1.3:

	0	a	b	c
0	0	0	0	0
a	0	a	b	c
b	0	b	0	b
c	0	c	b	a

Table 1.4:

	0	a	b	c
0	0	0	0	0
a	0	a	b	c
b	0	b	b	0
c	0	c	0	c

This is a ring isomorphic with the usual direct product of rings $\mathbb{Z}_2 \times \mathbb{Z}_2$. This ring is not isomorphic with the above one (isomorphic rings must have the same number (this being finite) of zero-square elements).

Case 4: $b.b = c$. We obtain the only field with four elements (denoted in the sequel $\mathcal{K}_4$) given by the multiplication in table 1.5

Table 1.5:

	0	a	b	c
0	0	0	0	0
a	0	a	b	c
b	0	b	c	a
c	0	c	a	b

Indeed $b^{-1} = c, c^{-1} = b$.

Remark. If two rings are isomorphic, their subjacent additive groups are also (group) isomorphic. Hence finally we have four nonisomorphic rings with identity.

Ex. 1.10 We first prove that all the ring structures one can define on a cyclic (abelian) group are commutative (i.e. commutativity follows from the distributivity laws). Indeed, $\bar{m} * \bar{n} = (\bar{1} + .. + \bar{1}) * (\bar{1} + .. + \bar{1}) = \bar{1} * \bar{1} + .. + \bar{1} * \bar{1} = \bar{n} * \bar{m}$ because we use only the commutativity of integer multiplication. In what follows we use the notation of the previous exercise. We begin with three remarks:

(i) if the product of two nonzero elements of a, b, c is commutative then the corresponding ring is commutative. Indeed, if for instance $a.b = b.a$ then , $a.c = a(a + b) = a.a + a.b = a.a + b.a = (a + b)a = c.a$; the equality $b.c = c.b$ follows analogously.

(ii) if $a.a = b$ the corresponding ring is commutative. Indeed, $a.b = a.(a.a) = (a.a).a = b.a$ follows from associativity and we apply (i). Hence $a.a = 0$ or $a.a = a$ (because the case $a.a = c$ is similar) so that in order to construct non-commutative rings on a set with four elements one has to take only idempotent or zero-square elements.

(iii) On $\mathcal{K}$, the products $\{a.a, b.b, c.c, a.b\}$ determine the ring multiplication. Indeed: $b.a = (c + a)a = c.a + a.a = c(c + b) + a.a = c.c + (b + a)b + a.a = a.a + b.b + c.c + a.b, a.c = a.a + a.b, c.a = b.b + c.c + a.b, b.c = a.a + c.c + a.b, c.b = b.b + a.b$. So we have to distinguish the following four cases:

Case 1 $a.a = b.b = c.c = 0$. The above computations show that $a.b = b.a$. Hence the corresponding ring is commutative.

Case 2 $a.a = a, b.b = c.c = 0$. In these conditions $b.a = a + a.b = b.c$ so that (using $x + x = 0$ for each $x \in \mathcal{K}$) we deduce a contradiction as follows: $a = a + (ab + ab) = ba + ab = ba + (a + ab)b = baa + bab = ba(a + b) = bac = bcc = 0$. Hence there are no ring structures in this case.

Case 3 $a.a = a, b.b = b, c.c = 0$. If moreover $a.b = 0$ we obtain $b.c = a$ which leads to a contradiction, $a = 0$, by left multiplication with a. If $a.b = c$ we derive a contradiction, $a = 0$, via $c.b = a$ by right multiplication with a (using also $b.a = 0$). The remaining cases are $a.b \in \{a, b\}$. We obtain on $\mathbb{Z}_2 \times \mathbb{Z}_2$ the following two non-commutative multiplications (see table 1.6 and 1.7)

Table 1.6:

	0	a	b	c
0	0	0	0	0
a	0	a	a	0
b	0	b	b	0
c	0	c	c	0

and

Table 1.7:

	0	a	b	c
0	0	0	0	0
a	0	a	b	c
b	0	a	b	c
c	0	0	0	0

that is two nonisomorphic rings C_4 and D_4. If $f : C_4 \to D_4$ would be a ring isomorphism, first one shows that $f(a) \in \{0, a, c\}$ lead to contradictions. The remaining case $f(a) = b, f(b) = a$ implies $b = f(a) = f(a.b) = f(a).f(b) = b.a = a$, a contradiction.

Case 4 $a.a = a, b.b = b, c.c = c$. We obtain at once $b.a = a.b$ and hence the corresponding rings are commutative. Finally, there are only two nonisomorphic non-commutative ring structures on a set with four elements (obviously each such ring is the *opposite* ring for the other ring).

Ex. 1.11 The only (abelian) group with p elements being $\mathbb{Z}_p$, two ring structures on this group are already well-known: $(\mathbb{Z}_p, +, \circ)$ with $\overline{n} \circ \overline{m} = \overline{0}$ for each $\overline{n}, \overline{m} \in \mathbb{Z}_p$ (i.e. the zero-square structure) and the usual structure $(\mathbb{Z}_p, +, \cdot)$. In what follows we prove that if A is a ring structure on $\mathbb{Z}_p$ which is not the zero-square structure, A is isomorphic to the usual structure. If $A^2 \neq \{0\}$ there are elements $a, b \in A$ such that

$ab \neq 0$. The subset $B = \{x \in A | ax = 0\}$ is a proper right ideal of A so that (being a subgroup of a simple group) $B = \{0\}$ and hence a is no left zero divisor in A. Now, $a \notin C = \{x \in A | xa = 0\}$ and as above $C = \{0\}$ so that a is right zero divisor neither. $(A, \cdot)$ being a finite semigroup and a being a cancelable element, a is an identity in A. But hence the function $F : \mathbb{Z} \to A$, $F(n) = nu, \forall n \in \mathbb{Z}$ is a surjective ring homomorphism with kernel $p\mathbb{Z}$ so that A is isomorphic with the usual structure.

Ex. 1.12 The map $\varphi : R \to End(R, +)$, $\varphi(a) = t_a$, $\forall a \in R$ where $t_a : R \to R$, $t_a(x) = a \cdot x$ is easily checked to be an embedding (i.e. injective homomorphism of rings which preserves the identities).

The condition (a) is equivalent to the surjectivity of φ. If $R = \mathbb{Z}$ and $x \in \mathbb{N}^*$ then for each $f \in End(\mathbb{Z}, +)$ we have $f(x) = f(1 + .. + 1) = f(1) + .. + f(1) = f(1).x$ and $f(0) = 0$, $f(-x) = -f(x)$ so that $(\mathbb{Z}, +, \cdot) \cong (End(\mathbb{Z}, +), +, \circ)$. If $R = \mathbb{Q}$ and $f \in End(\mathbb{Q}, +)$ one has $f(1) = f(\frac{n}{n}) = f(\frac{1}{n} + .. + \frac{1}{n}) = f(1).\frac{1}{n}$ and hence $f(\frac{m}{n}) = mf(\frac{1}{n}) = \frac{m}{n} \cdot f(1)$ for every $\frac{m}{n} \in \mathbb{Q}$ so that $(\mathbb{Q}, +, .) \cong End(\mathbb{Q}, +), +, \circ)$.

(b) The sufficiency of the condition being immediate, conversely, if $(End(R, +)$ is commutative $f(x) = f(x \cdot 1) = f(t_x(1)) = (f \circ t_x)(1) = (t_x \circ f)(1) = t_x(f(1)) = x \cdot f(1)$ so by (a) the condition is also necessary.

Ex. 1.13 (a) Let $b \in R$ be a left inverse of a and $c = 1 - ab$. We check that $b + c$ is another left inverse of a. Indeed, $c \neq 0$ because otherwise b would be a right inverse for a. So $b + c \neq b$. Moreover, $(b + c)a = (b + 1 - ab)a = 1 + a - a = 1$.

(b) Consider $A = \{b \in R \mid ba = 1\}$ where $A \neq \emptyset$ and let $a_0 \in A$ fixed. The function $f : A \to A$ defined by $f(b) = ab - 1 + a_0$ is injective but not surjective: $f(b) = f(c) \Rightarrow ab = ac \Rightarrow b = c$ (by left multiplication with b) and $a_0 \notin f(A)$ (if a_0 would be an image $f(b)$ then $ab = 1$ and a (being a unit) would have a unique (left-sided) inverse, contradicting the hypothesis:A has at least two elements). Finally, for a finite set X and $\alpha : X \to X$, the properties "α injective" and "α surjective" are equivalent. Hence A is infinite.

Ex. 1.14 (a) If $1 \in R$ then $\mathcal{M}_n(R)$ has the well-known identity matrix

$$I_n = \begin{pmatrix} 1 & 0 & . & . & 0 \\ 0 & 1 & . & . & 0 \\ . & . & . & . & . \\ . & . & . & . & . \\ 0 & 0 & . & . & 1 \end{pmatrix}$$. Conversely, if $E = (e_{ij})_{1 \le i, j \le n}$ is the identity of

$\mathcal{M}_n(R)$ and $a \in R$, using the matrices $A = (a_{ij})$ with all zero entries but $a_{11} = a$ and the equalities $EA = AE$, one proves that e_{11} is an identity in R.

(b) If $R^2 = \{0\}$ then $(\mathcal{M}_n(R))^2 = \{0_n\}$ and $\mathcal{M}_n(R)$ obviously is a commutative ring. Conversely, if $\mathcal{M}_n(R)$ is commutative and $a, b \in R$

we get $a.b = 0$ from $\begin{pmatrix} a & 0 & . & . & 0 \\ 0 & 0 & . & . & 0 \\ . & . & . & . & . \\ . & . & . & . & . \\ 0 & 0 & . & . & 0 \end{pmatrix} \begin{pmatrix} 0 & . & . & 0 & b \\ 0 & . & . & 0 & 0 \\ . & . & . & . & . \\ . & . & . & . & . \\ 0 & . & . & 0 & 0 \end{pmatrix} =$

$\begin{pmatrix} 0 & . & . & 0 & b \\ 0 & . & . & 0 & 0 \\ . & . & . & . & . \\ . & . & . & . & . \\ 0 & . & . & 0 & 0 \end{pmatrix} \begin{pmatrix} a & 0 & . & . & 0 \\ 0 & 0 & . & . & 0 \\ . & . & . & . & . \\ . & . & . & . & . \\ 0 & 0 & . & . & 0 \end{pmatrix}$.

(c) If $r \in Z(R)$ and $A \in \mathcal{M}_n(R)$ then $rI_n \cdot A = rA = Ar = A \cdot rI_n$. Conversely, if $U = (u_{ij}) \in Z(\mathcal{M}_n(R))$ let us consider a matrix E_{ij} (a so called *matric unit*) with 1 on the i-th row and j-th column and all the other entries zero. From the equalities $UE_{ij} = E_{ij}U$ we deduce $u_{ii} = u_{jj}$ for each $i, j \in \{1, 2, .., n\}$ and $u_{ij} = 0$ for $i \ne j$ so that U is diagonal, say $u \cdot I_n$. Now if $r \in R$ from $(u \cdot I_n)(r \cdot E_{11}) = (r \cdot E_{11})(u \cdot I_n)$ we also infer $u \in Z(R)$.

Ex. 1.15 If $e^2 = e \in R$ is idempotent and $a \in R$ easy computations lead to $(ea - eae)^2 = 0$ and $(ae - eae)^2 = 0$. Hence by hypothesis $ea = eae = ae$ holds and $e \in Z(R)$.

Ex. 1.16 The elements $a^2 - a, b^2 - b, (a+b)^2 - (a+b) = (a^2 - a) + (b^2 - b) + ab + ba$ belong to $Z(R)$ so that $ab + ba \in Z(R)$. Then $a(ab + ba) = (ab + ba)a$ or $a^2 b = ba^2$. These two elements being arbitrary $a^2 \in Z(R)$ and hence $a = a^2 - (a^2 - a) \in Z(R)$.

Ex. 1.17 If $(a + bi)(c + di) = 1$ then (computing the absolute values) $(a^2 + b^2)(c^2 + d^2) = 1$. The last product having terms in $\mathbb{Z}$ we must have $a^2 + b^2 = 1$ so that $\{1, -1, i, -i\}$ are the only units in $\mathbb{Z}[i]$. This is a cyclic group of order four.

Ex. 1.18 Let $N : \mathbb{Z}[\sqrt{d}] \to \mathbb{Z}$ be a map defined as follows: $N(a + b\sqrt{d}) = a^2 - db^2$ for each square-free integer d. For $x = a + b\sqrt{d}$ we denote by $\bar{x} = a - b\sqrt{d}$. Then the following hold: $N(x) = x.\bar{x} = N(\bar{x}), \overline{x.y} = \bar{x}.\bar{y}, N(x.y) = N(x).N(y)$. If x is a unit in $\mathbb{Z}[\sqrt{d}]$ then, for a suitable $y \in \mathbb{Z}[\sqrt{d}]$, we have $1 = N(1) = N(xy) = N(x)N(y)$ in $\mathbb{Z}$ so that $N(x) \in \{-1, 1\}$. Conversely, if $N(x) \in \{-1, 1\}$ then $x \cdot (\pm\bar{x}) = 1$ and x is a unit in $\mathbb{Z}[\sqrt{d}]$. Now, for $d = 2$ the real number $a = 1 + \sqrt{2}$ is a unit of $\mathbb{Z}[\sqrt{d}]$ because $N(a) = -1$. We check immediately that all $a^n (n \in \mathbb{N})$ are different units in $\mathbb{Z}[\sqrt{2}]$ (a strictly increasing sequence of limit ∞), e.g. $3 + 2\sqrt{2}, 7 + 5\sqrt{2}, ..$

Ex. 1.19 Let $u \in R$ be an inverse for $1 - ab$. Then one easily verifies that $1 + bua$ is the inverse of $1 - ba$. Indeed, for instance $(1 - ba)(1 + bua) = 1 - ba + bua - babua = 1 - ba + b(1 - ab)ua = 1 - ba + ba = 1$.

Ex. 1.20 If $n = p^2 t$ with p prime then $\bar{0} \neq \overline{pt}$ has zero square (is nilpotent). Conversely, let n be a square-free integer, i.e. $n = p_1 p_2 ... p_k$ and $\bar{a}^m = \bar{0}$ a nilpotent element. We prove that $\bar{a} = \bar{0}$. Indeed, $\bar{a}^m = \bar{0}$ in $\mathbb{Z}_n$ implies $n | a^m$ and hence $p_i | a^m$ ($i \in \{1, 2, .., k\}$). Then $p_i | a$ and $n | a$ (because p_i are different primes) so that $\bar{a} = \bar{0}$.

Ex. 1.21 If $n = p_1^{r_1}..p_k^{r_k}$ and $\bar{m}$ is nilpotent in $\mathbb{Z}_n$ let $(\bar{m})^t = \bar{0}$. Then n divides m^t, and so each p_i divides m^t. The numbers $\{p_i | 1 \leq i \leq k\}$ being prime, each p_i divides m. Being different primes to each other, even the product $p_1..p_k$ divides m. Conversely, let $m = sp_1..p_k$ be an integer with $s \in \mathbb{Z}$ and let $r = \max_{1 \leq i \leq k} r_i$. Obviously, $m^r = s^r p_1^r..p_k^r = n.s^r p_1^{r-r_1}..p_k^{r-r_k}$ and hence $(\bar{m})^r = \bar{0}$, i.e. $\bar{m}$ is nilpotent in $\mathbb{Z}_n$.

Ex. 1.22 Let $n = p_1^{r_1} p_2^{r_2}..p_k^{r_k}$ the prime decomposition of n and let A be the set of all the subsets of $\{p_1, p_2, .., p_k\}$. If $Id(\mathbb{Z}_n)$ denotes the set of all the idempotent elements of $\mathbb{Z}_n$, in what follows we will prove that the function $\varphi : Id(\mathbb{Z}_n) \to A$ defined by $\varphi(\bar{x}) = \{p \in A | p$ divides $(x; n)\}$ (as above $(x; n)$ denotes the g.c.d.) for every $\bar{x} \in \mathbb{Z}_n$, is bijective. From this proof we also infer a method of construction for all the idempotent elements of $\mathbb{Z}_n$.

We first prove that $\varphi(\bar{x}) = \varphi(\bar{y})$ implies (for idempotent elements) $(x; n) = (y; n)$ property stronger than "φ is well-defined" (i.e. $\bar{x} = \bar{y} \Rightarrow$ $(x; n)$ and $(y; n)$ have the same prime divisors). Let $\bar{x}$ be an idempotent element in $\mathbb{Z}_n$: $\bar{x}^2 = \bar{x}$ or $\bar{x}(\bar{x} - \bar{1}) = \bar{0}$ or $n|x(x-1)$. If $d = (x; n)$, $x = x'd$, $n = n'd$ we surely have $(x'; n') = 1$. From $x(x-1) = tn$ we get $n'|x'(x-1)$ and hence $n'|x-1$. In general x and $x-1$ are relatively prime so that d and n' (or their divisors) are also relatively prime. Hence, if $p_i|d$ then $p_i^{r_i}|d$ so that $\varphi(\bar{x})$ characterizes $(x; n)$, i.e. $\varphi(\bar{x}) = \varphi(\bar{y}) \Rightarrow (x; n) = (y; n)$.

Next we prove that φ is injective. If $\varphi(\bar{x}) = \varphi(\bar{y})$ using the above arguments $(x; n) = (y; n)$ and $n' = \frac{n}{(x;n)} = \frac{n}{(y;n)}$ divides $x-1$ respectively $y-1$ and hence $x-y$. Clearly $(x; n) = (y; n)$ also divides $x-y$ and being relatively prime with n' we obtain $n = dn'|x-y$. Hence $\bar{x} = \bar{y}$.

Finally, in order to verify that φ is surjective, let $\{p_i\}_{i \in I} \in A$ and $a = \prod_{i \in I} p_i^{r_i}$ (with $a = 1$ if $I = \emptyset$). Clearly a and $\frac{n}{a}$ have no common divisors so that there are $u, v \in \mathbb{Z}$ such that $ua + v(\frac{n}{a}) = 1$. We choose $x = ua$ and then $x(x-1) = ua.(-v\frac{n}{a}) = -uvn$ or $n|x(x-1)$. Then $\bar{x} \in Id(\mathbb{Z}_n)$ and from $(x; n) = a$ we get $\varphi(\bar{x}) = \{p_i\}_{i \in I}$.

Examples: $n = 12 = 2^2.3$, $A = \{\emptyset, \{2\}, \{3\}, \{2, 3\}\}$ so that with the above notations $a \in \{1, 4, 3, 12\}$ and $u \in \{1, 1, -1, 0\}$. We find $Id(\mathbb{Z}_{12}) = \{\bar{1}, \bar{4}, \bar{9}, \bar{0}\}$.

$n = 360 = 2^3.3^2.5$; $a \in \{1, 8, 9, 5, 72, 40, 45, 360\}$ and $Id(\mathbb{Z}_{360}) = \{\bar{1}, \overline{136}, \overline{81}, \overline{145}, \overline{216}, \overline{280}, \overline{225}, \bar{0}\}$.

The application follows immediately using the bijection φ defined above.

Ex. 1.23 (a) One verifies that $a + a = 0$ for each $a \in R$ (see 5.1). Hence $a + b = (a + b)^2 = a^2 + ab + ba + b^2 = a + b + ab + ba$ so that $ab + ba = 0$ or $ab = -ba = ba$ (because also $ba + ba = 0$).

(b) If R is a Boole ring then $a = a^2 = .. = a^n = ..$ so that obviously zero is the only nilpotent element. Moreover, $(a + b)ab = a^2b + ab^2 = ab + ab = 0$. Conversely, we first take $b = a$ in the relation given in the hypothesis: $(a + a)aa = 0$ implies $a^3 + a^3 = 0$ so that (by succesivess multiplication with a) $a^n + a^n = 0$ for each $n \in \mathbb{N}, n \geq 3$.

Further, let $b = a^2 - a$; we obtain $a^4 = a^5$ and so (again by succesivess multiplication with a) $a^4 = a^5 = .. = a^n = ..$ for each $n \in \mathbb{N}, n \geq 4$. In particular $a^8 = a^4$ so that we shall verify $(a^2 - a)^4 = 0$. Indeed, $(a^2 - a) = a^4 + a^2 - a^3 - a^3 = a^4 + a^2$ and hence $(a^2 - a)^4 = (a^4 + a^2)^2 = a^8 + a^6 + a^6 + a^4 = a^4 + a^4 = 0$. Finally, R having no nonzero nilpotent elements, $a^2 - a = 0$ and $a^2 = a$.

Remark. If R has identity the solution is immediate (taking $b = 1$ and $b = a$) and the condition on nilpotent elements is no more necessary.

Ex. 1.24 From $a = xab$ we infer $a = a^2 = (xab)^2 = x(xa^2b^2) = x(xab) = xa$ respectively from $x = a + b + ab$ we compute $xa = (a + b + ab)a = a^2 + ab + a^2b = a^2 + ab + ab = a^2 = a$ the multiplication (see the previous exercise) being commutative and the characteristics (for definition see chapter 5) being 2. Finally, the required element is

$$x = \sum_{i=1}^{n} a_i + \sum_{1 \leq i < j \leq n} a_i a_j + .. + a_1 a_2 .. a_n.$$

Ex. 1.25 If $yz = 1$ and $x^n = 0$, in a commutative ring, a simple computation leads to $(y + x)(x - xz^2 + x^2z^3 - .. + (-1)^{n-1}x^{n-1}z^n) = yz + (-1)^n x^n z^n = yz = 1$. If $y = 1$ the commutativity of R is not necessary: $(1 + x)(1 - x + x^2 - .. + (-1)^{n-1}x^{n-1}) = 1$. In $M_2(\mathbb{Z})$ consider $A = \begin{pmatrix} 0 & 1 \\ 1 & 0 \end{pmatrix}, B = \begin{pmatrix} 0 & -1 \\ 0 & 0 \end{pmatrix}$. Simple verifications lead to

$AB \neq BA$, A unit, B nilpotent and $A + B = \begin{pmatrix} 0 & 0 \\ 1 & 0 \end{pmatrix}$ is not a unit.

Ex. 1.26 Obviously, T is a subgroup of $(\mathcal{M}_n(R), +)$. If $A = (a_{ij})$, $B = (b_{ij})$ and $A \cdot B = C = (c_{ij})$ then $c_{ij} = \sum_{k=1}^{n} a_{ik}b_{kj} = \sum_{k<i} a_{ik}b_{kj} + \sum_{k=i}^{n} a_{ik}b_{kj} = 0 + \sum_{k=i}^{n} a_{ik}b_{kj}$. Now if $i > j$ then $k \geq i > j$ implies $b_{kj} = 0$ and so $c_{ij} = 0$.

Ex. 1.27 If the given subset would be a subring and $u = \sqrt[3]{5}$, integers a, b would exist such that $u^2 = au + b$. Hence $5 = u^3 = u(au + b) = a(au + b) + bu = (a^2 + b)u + ab$ so that the system $a^2 + b = 0 = ab - 5$ would have solutions in $\mathbb{Z}$. But this is clearly not the case and the subset is no subring of $\mathbb{C}$.

Ex. 1.28 One has clearly $A = \langle(\bar{1},\bar{1})\rangle = \left\{(\bar{0},\bar{0}),(\bar{1},\bar{1}),(\bar{2},\bar{2}),(\bar{3},\bar{3})\right\}$ (the subring with identity generated by the identity) and $\mathbb{Z}_4 \times \mathbb{Z}_4$ as subrings with identity. Each subring with identity contains A so let B be such a subring and $(\bar{a},\bar{a}+\bar{b}) \in B$ with $\bar{b} \neq \bar{0}$. If $\bar{b} \in \left\{\bar{1},\bar{3}\right\}$ then B is the entire ring and if $\bar{b} = \bar{2}$ then $B = A \cup \left\{(\bar{2},\bar{0}),(\bar{0},\bar{2}),(\bar{1},\bar{3}),(\bar{3},\bar{1})\right\}$.

Ex. 1.29 Only the second part needs concern. If $a^2 + b^2 \neq 0$ then the determinant $\begin{vmatrix} a & b \\ qb & a \end{vmatrix} = a^2 - qb^2 \neq 0$ iff q is no square of a rational number (the matrices with $b = 0$ but $a \neq 0$ have obvious inverses).

Ex. 1.30 For $P \in \mathbb{P}$ we consider $A(P) = \left\{\frac{m}{n} \in \mathbb{Q} | p|n \Rightarrow p \in P\right\}$ that is, all the (irreducible) fractions $\frac{m}{n}$ with the prime divisors of n only in P and, for a subring A with identity of $\mathbb{Q}$ we consider $P(A)$ the set of all the prime divisors of the denominators of all the (irreducible) fractions from A. We shall prove that these two correspondences are inverse one to another (i.e. $A(P(A)) = A$ and $P(A(P)) = P$) so that there is a bijection between the set of all the subrings with identity of $\mathbb{Q}$ and $\mathcal{P}(\mathbb{P})$ the set of all the (finite or infinite) subsets of prime numbers.

We begin by several elementary remarks. All the fractions are supposed to be irreducible. If a subring with identity contains $\frac{m}{n}$ it must also contain $\frac{1}{n}$. Indeed, if $(m;n) = 1$ then $um + vn = 1$ holds for suitable $u,v \in \mathbb{Z}$ and hence $\frac{1}{n} = \frac{um+vn}{n} = u\frac{m}{n} + v \cdot 1$. Further, $\mathbb{Z} = \langle 1 \rangle$ (the subring generated by 1) is clearly the smallest subring with identity of $\mathbb{Q}$. Finally, $P(A(P)) = P$ and $A \subseteq A(P(A))$ are immediate by the above definitions. In order to verify $A(P(A)) \subseteq A$ let $\alpha = \frac{m}{p_1^{r_1}..p_k^{r_k}} \in A(P(A))$; from $p_1 \in P(A)$ there is a fraction $\frac{s}{p_1^{t}t} \in A$ and by an above remark $\frac{1}{p_1} \in A$. Similarly, $\frac{1}{p_2},..,\frac{1}{p_k} \in A$ so that $\alpha \in A$.

Remark. If $P = \{p_1, ..p_k\}$ is finite then $P(A)$ is the subring generated by $\frac{1}{p_1..p_k}$.

Ex. 1.31 From $x^6 = x$ and $(-x)^6 = -x$ we infer $x = -x$ and so $char(R) = 2$. From $(x + 1)^6 = x + 1$ developing the left member of this equality (notice that no commutativity is required) we obtain $(x^6 - x) + 2(3x^5 + 7x^4 + 10x^3 + 7x^2 + 3x) + (x^4 + x^2) = 0$ or $x^4 + x^2 = 0$. Hence $x^4 = -x^2 = x^2$ and, multiplying by x^2, we finally have $x = x^6 = x^4 = x^2$.

Chapter 2

Ideals

Ex. 2.1 Only P and A are subrings. A is also an ideal but P is not:
if $f \in P$ and $g \in F, g(x) = \begin{cases} x \sin \frac{1}{x} & \text{if } x \neq 0 \\ 0 & \text{if } x = 0 \end{cases}$
then $f \cdot g$ has an infinite number of zero's and hence $f \cdot g \notin P$.

Ex. 2.2 As a Boole ring, R is commutative with identity (see 1.23) so that the principal ideals are $(A) = A.R = \mathcal{P}(A)$ and $(B) = B.R = \left\{\emptyset, \left\{\frac{1}{4}\right\}\right\}$. Further, from $\frac{1}{4} \in \left[0, \frac{1}{2}\right]$ we infer $(A) \circ (B) = (B)$. Finally, $(A, B) = (A) + (B) = \mathcal{P}(A) + \left\{\emptyset, \left\{\frac{1}{4}\right\}\right\} = \mathcal{P}(A) = (A)$.

Ex. 2.3 (a) Obvious.

(b) For instance, if X is a left ideal, $r \in R$ and $a \in l(X)$ then $(ar)x = a(rx) = 0$ for each $x \in X$ because $rx \in X$. Hence $l(X)$ is a right ideal (and so twosided according to (a)).

(c) The inclusions (i) are straightforward. (ii) Let $x \in X$; for every $a \in l(X) : ax = 0$. But this implies $x \in r(l(X))$. The second inclusion is similar. (iii) From (i) and (ii) $X \subseteq r(l(X)) \Rightarrow l(X) \supseteq l(r(l(X)))$. Replacing X by $l(X)$ and using again (ii) we have $l(X) \subseteq l(r(l(X)))$ and the required equality. The last equality is analoguous.

Ex. 2.4 If e is a left identity modulo I and e' a left identity modulo J then let $e = x_1 + y_1, e' = x_2 + y_2 (x_1, x_2 \in I, y_1, y_2 \in J)$. We prove that $\bar{e} = x_2 + y_1$ is a left identity modulo $I \cap J$. Together with $x_1 = e - y_1$ the element $u = (e - y_1)r - x_2r \in I$ for each $r \in R$. But

91

$u = r - (x_2 + y_1) r - (r - er)$ and so, together with $r - er$ the element $r - (x_2 + y_1) r \in I$. The rest follows symmetrically.

Ex. 2.5 Generally, if $x \in R$, a commutative ring, then $(x) = Rx + \mathbb{Z}x$. In particular, for $R = n\mathbb{Z} \subseteq \mathbb{Z}$ we get $(x) = \mathbb{Z}x, \forall x \in n\mathbb{Z}$. Denoting by $\mathcal{S}(n\mathbb{Z})$ the set of all the (cyclic) subgroups of the infinite cyclic group$(\mathbb{Z}, +)$ and by $\mathcal{I}(n\mathbb{Z})$ the set of all the ideals of $n\mathbb{Z}$ one has $\mathcal{S}(n\mathbb{Z}) = \{mn\mathbb{Z} | m \in \mathbb{Z}\} = \{(x) | x \in n\mathbb{Z}\} \subseteq \mathcal{I}(n\mathbb{Z}) \subseteq \mathcal{S}(n\mathbb{Z})$ and hence all the inclusions must be equalities. So $n\mathbb{Z}$ is a principal ring.

Ex. 2.6 The inclusion $I + (J \cap K) \subseteq (I + J) \cap K$ is clearly always valid if $I \subseteq K$. Conversely, if $k \in (I + J) \cap K, k = i + j$ $(i \in I, j \in J, k \in K)$ then $j = k - i \in K$ because $I \subseteq K$ and hence $k = i + j \in I + (J \cap K)$.

Ex. 2.7 $2\mathbb{Z} \cup 3\mathbb{Z}$ is not an ideal in $(\mathbb{Z}, +, .)$ neither a subgroup of $(\mathbb{Z}, +) : 2 + 3 = 5 \notin 2\mathbb{Z} \cup 3\mathbb{Z}$.

Ex. 2.8 Consider the abelian group $(\mathbb{Q}, +)$ together with the zero multiplication $\frac{a}{b} \cdot \frac{c}{d} = 0, \forall \frac{a}{b}, \frac{c}{d} \in \mathbb{Q}$. The ideals in this ring are precisely the subgroups of $(\mathbb{Q}, +)$. But, being divisible, $\mathbb{Q}$ has no maximal subgroups (or: each proper subgroup H has infinite index, so that $\mathbb{Q}/H$ is not simple) and hence $\mathbb{Z}$ is an ideal not contained in any maximal ideal.

Ex. 2.9 One verifies immediately that $a^{-1}M = \{r \in R | ar \in M\}$ is a right ideal in a ring R. Consider R/M and $R/a^{-1}M$ as right R-modules (R has identity) and the following right R-module homomorphism $f :$ $R \to R/M, f(r) = ar + M, \forall r \in R$. It is easy to observe that R/M is simple (irreducible) as right R-module iff M is a right maximal ideal. These two right R-modules are isomorphic: one uses a well-known isomorphism theorem $R/M = im(f) \cong R/\ker(f)$ because $\ker(f) = a^{-1}M$ and f is surjective (indeed, $a \notin M \Rightarrow aR \not\subseteq M \Rightarrow M \subset M + aR \Rightarrow M + aR = R$ proves that $\forall b + M \in R/M, \exists r \in R : b + M = f(r)$). Hence $R/a^{-1}M$ is also simple and $a^{-1}M$ is a right maximal ideal.

Ex. 2.10 An obvious induction reduces this to a 2-generated ideal, i.e. $A = Rx_1 + Rx_2$. Each element being idempotent (this assures that the ring is commutative), one easily verifies the equalities: $x_1 = x_1(x_1 + x_2 - x_1 x_2)$ and $x_2 = x_2(x_1 + x_2 - x_1 x_2)$. Hence $A = Rx_1 + Rx_2 = R(x_1 + x_2 - x_1 x_2)$.

Ex. 2.11 If for two (commutative) rings with identity R and S we have I and J respectively ideals then it is easy to check that $I \times J$ is an ideal of $R \times S$. Conversely, if A is an ideal of $R \times S$ then denoting by p_R and p_S the projections from the direct product one verifies that $A = p_R(A) \times p_S(A)$ where $p_R(A)$ is ideal in R and $p_S(A)$ is ideal in S. (The astute reader will observe that the identities of R and S are needed!). Now every division ring K has only the trivial ideals $\{0\}$ and K. Hence $\{0\} \times K$ and $K \times \{0\}$ are the only nontrivial ideals of $K \times K$.

Moreover, the ideals of $\mathbb{Z} \times \mathbb{Z}$ have the form $m\mathbb{Z} \times n\mathbb{Z}$ for $m, n \in \mathbb{N}$. Finally, let us remark that this fails for subrings in a direct product: $\Delta_R = \{(r,r)|r \in R\}$ is a subring of $R \times R$ which has not the decomposition $A \times B$ for subrings A, B of R.

Ex. 2.12 Let R be a nonzero ring with identity. The set A of all the matrices with zero entries on the first column form a left ideal of $\mathcal{M}_n(R)$ (which is not a right ideal) and the set B of all the matrices with zero entries on the first row form a right ideal of $\mathcal{M}_n(R)$ (which is not a left ideal). The intersection $A \cap B$ is no ideal of $\mathcal{M}_n(R)$ because for instance

$$\begin{pmatrix} 0 & 0 & . & . & 0 \\ 0 & 1 & . & . & 1 \\ . & . & . & . & . \\ . & . & . & . & . \\ 0 & 1 & . & . & 1 \end{pmatrix} \begin{pmatrix} 1 & 1 & . & . & 1 \\ 1 & 1 & . & . & 1 \\ . & . & . & . & . \\ . & . & . & . & . \\ 1 & 1 & . & . & 1 \end{pmatrix} =$$

$$\begin{pmatrix} 0 & 0 & . & . & 0 \\ n-1 & n-1 & . & . & n-1 \\ . & . & . & . & . \\ . & . & . & . & . \\ n-1 & n-1 & . & . & n-1 \end{pmatrix} \notin A \text{ and hence } \notin A \cap B.$$

Remark. If R is a division ring (or a simple ring) then $\mathcal{M}_n(R)$ is simple (see also 2.20) and the above example (for $n \geq 2$) is simplier to justify: $A \cap B \notin \{0, R\}$.

Ex. 2.13 .(a) $r_1, r_2 \in A : B$ implies $\forall b \in B \ (r_1 - r_2)b = r_1 b - r_2 b \in A$ (A being subgroup of $(R, +)$) so that $A : B$ is a subgroup of $(R, +)$. For an arbitrary $c \in R$ and $r \in A : B$ the element $cr \in A : B$ because $\forall b \in B \ rb \in A \Rightarrow (cr)b = c(rb) \in A$ (A ideal) and $rc \in A : B$ because $\forall b \in B \ cb \in B$ (B being ideal) so that $(rc)b = r(cb) \in A$.

(b) The inclusion $(A : B).B \subseteq A$ being obvious let I be an ideal such that $(A : B).I \subseteq A$. One shows that $I \subseteq A : B$ and that between all the upper bounds $A : B$ is the lower.

(c) The first and second equality are immediate. As for the third, $r \in A : (B_1 + .. + B_n)$ implies $\forall b_i \in B_i$ $(i \in \{1, .., n\}) : r(b_1 + .. + b_n) \in A$. Taking for example $b_i = 0$ $(i \in \{2, .., n\})$ we get $rb_1 \in A$ or $r \in A : B_1$. The rest is analogous; the converse inclusion is simple.

(d) Owing to the well-known form of the ideals in $\mathbb{Z}$, the following holds $n\mathbb{Z} : m\mathbb{Z} = \{a \in \mathbb{Z} | am\mathbb{Z} \subseteq n\mathbb{Z}\}$. But

$$am\mathbb{Z} \subseteq n\mathbb{Z} \Leftrightarrow n | am \Leftrightarrow \tfrac{[m;n].(m;n)}{m} | a \tfrac{[m;n].(m;n)}{n} \Leftrightarrow \tfrac{[m;n]}{m} | a$$

because $\tfrac{[m;n]}{m}$ and $\tfrac{[m;n]}{n}$ are relatively prime. Hence $n\mathbb{Z} : m\mathbb{Z} = \tfrac{[m;n]}{m}\mathbb{Z}$.

Ex. 2.14 First observe that for 3 ideals I, J, K in R one has $J \circ K \subseteq I \Leftrightarrow K \subseteq I : J$. Indeed, if $k \in K$ and $J \circ K \subseteq I$ then $kJ \subseteq I$ and hence $k \in I : J$. Conversely, from $K \subseteq I : J$ one has $kJ \subseteq I$ for every $k \in K$ and so $K \circ J \subseteq I$. Notice that in this exercise the ring R is commutative and hence $I \circ J = J \circ I$.

(i) From $A : B \subseteq A : B$ we obtain (using the above remark) $B \circ (A : B) = (A : B) \circ B \subseteq A$. As for the second, from the previous exercise $A : (A + B) = (A : A) \cap (A : B) = R \cap (A : B) = A : B$.

(ii) If $r \in A : (B \circ C)$ then $rB \circ C \subseteq A$ and so $rC \subseteq A : B$. Hence $r \in (A : B) : C$ and all the rest is similar.

(iii) Clearly, $rB \subseteq B$ for each $r \in R$ so that $B \subseteq A$ implies $A : B = R$. Conversely, if R has identity, $1 \in A : B$ and so $1.B \in A$ or $B \subseteq A$.

Ex. 2.15 (a)It is obvious that $r \in \sqrt{I}$ and $c \in R$ imply $rc, cr \in \sqrt{I}$ (R being commutative).For $r_1, r_2 \in \sqrt{I} \Rightarrow r_1 - r_2 \in \sqrt{I}$ one has to use the binomial formula (R commutative): $r_1^n, r_2^m \in I \Rightarrow (r_1 - r_2)^{n+m} =$

$$\binom{n+m}{0} r_1^{n+m} - \binom{n+m}{1} r_1^{n+m-1} r_2 + .. + (-1)^m \binom{n+m}{m} r_1^n r_2^m +$$

$$.. + (-1)^{n+m} \binom{n+m}{n+m} r_2^{n+m} ,$$ the members in "the left side" of the above sum being in I together with r_1^n and the members in "the right side" are in I together with r_2^m.

(b) The inclusion $I \subseteq \sqrt{I}$ being immediate, other three from the stated inclusions can be deduced. It remains: $r \in \sqrt{\sqrt{I}} \Rightarrow \exists n : r^n \in$

$\sqrt{I} \Rightarrow \exists m : (r^n)^m = r^{nm} \in I \Rightarrow r \in \sqrt{I}; \ r \in \sqrt{I} \cap \sqrt{J} \Rightarrow \exists n, m :$
$r^n \in I, r^m \in J \Rightarrow r^{n+m} = r^n.r^m \in I \cap J \Rightarrow r \in \sqrt{I \cap J}$: and
$r \in \sqrt{\sqrt{I} + \sqrt{J}} \Rightarrow \exists k : r^k = a + b, a \in \sqrt{I}, b \in \sqrt{J}$. In this last proof
if $a^n \in I$ and $b^m \in J$, we apply again the binomial formula for the
computation of $r^{k(n+m)} = (a+b)^{n+m} \in I+J$ and obtain the conclusion
$r \in \sqrt{I+J}$.

(c) Indeed, the set of all the nilpotent elements from R is clearly
$\sqrt{\{0\}}$, so it is an ideal in a commutative ring. The commutativity is
essential: in $\mathcal{M}_2(\mathbb{Z})$ the matrices $\begin{pmatrix} 0 & 1 \\ 0 & 0 \end{pmatrix}$ and $\begin{pmatrix} 0 & 0 \\ 1 & 0 \end{pmatrix}$ are nilpotent
(of zero square) but their sum $\begin{pmatrix} 0 & 1 \\ 1 & 0 \end{pmatrix}$ being a unit is not nilpotent.

Ex. 2.16 If I and J are nilideals (i.e. contain only nilpotent elements),
each element in $I + J$ is nilpotent too. Indeed, let $x + y = z \in I + J$
($x \in I, y \in J$) and $k \in \mathbb{N}^*$ such that $x^k = 0$. We have $z^k = (x+y)^k =$
$x^k + a$ with a suitable $a \in J$ (e.g. in a product $xyxy..xy$ each $xy \in J$
because J is a (left) ideal, $xyx \in J$ because J is a (right) ideal again,
etc.) Finally, there is a $l \in \mathbb{N}^*$ such that $a^l = 0$ and hence $z^{kl} = a^l = 0$
, z is nilpotent and $I + J$ is nil.

If I and J are nilpotent ideals and $I^n = J^m = 0$ then $(I + J)^{n+m} \subseteq$
$I^{n+m} + J^{n+m}$ shows that $I + J$ is nilpotent too. The verification of the
previous inclusion goes like this: any element from $(I+J)^{n+m}$ is a sum
of products which contain $n + m$ factors from I or J, that is (at least
n) factors from I or (at least m) factors from J (e.g. a finite product
of the form $a_1 b_1 a_2 b_2..$, where $a_i \in I, b_j \in J$ can be written as $a_1 a_2' a_3'..$
with $a_i' \in I$ -using I left ideal or $b_1' b_2'..$ with $b_j' \in J$ -using J left ideal)
and one uses the obvious inclusions $I^t \subseteq I^{n+m}, J^t \subseteq J^{n+m}$ for each
$t \leq n + m$.

Ex. 2.17 Obvious.

Ex. 2.18 All the subgroups of $\mathbb{Z}_{p^n}$ form a chain: $0 < p^{n-1}\mathbb{Z}_{p^n} <$
$.. < p\mathbb{Z}_{p^n} < \mathbb{Z}_{p^n}$ and these are the only ideals of $\mathbb{Z}_{p^n}$ too. These
are obviously nilpotent and $p\mathbb{Z}_{p^n}$ (which has index n) contains all the
nilpotent elements of $\mathbb{Z}_{p^n}$ -see also the solution of 1.20). Further, I is
clearly a nil ideal, each element having only a finite number of nonzero

components. I is not a nilpotent ideal because for each $m \in \mathbb{N}^*$ one can point out a nonzero product of $m + 1$ elements in I.

Ex. 2.19 Let $u \in y$ i.e. $u + U = y$. From $y^2 = y$ we infer $u^2 - u \in U$ and hence (U being nil) there is a $m \in \mathbb{N}^*$ such that $(u^2 - u)^m = 0$. We have $u^2 - u = u(1-u) = (1-u)u$ so that $0 = (u^2 - u)^m = u^m(1-u)^m = u^m - u^{m+1}.g(u)$ where g is a polynomial function with coefficients in R which commutes with all the powers of u. Now choose $x = u^m g(u)^m$. The following computations show that x is the required idempotent:
$x^2 = u^{2m} g(u)^{2m} = u^{m-1} u^{m+1} g(u) g(u)^{2m-1} = u^{m-1} u^m g(u)^{2m-1}$
$\quad = u^{2m-1} g(u)^{2m-1} = .. = u^m g(u)^m = x$, and $x + U = u^m g(u)^m + U = y(g(u)^m + U) = .. = y(g(u)^2 + U) = y(g(u) + U) = y$. Indeed, $y = u + U = u^2 + U = .. = u^m + U$ follows, y being idempotent and $y = y(g(u) + U)$ is deduced from $u^m = u^{m+1} g(u)$ and the previous equalities.

Ex. 2.20 Let K be a field and T the subring (see 1.26) of $\mathcal{M}_n(K)$ of all the triangular matrices. The subset A of all the matrices from T which have zero elements (also) on the diagonal (i.e. $a_{ii} = 0$ for each $i \in \{1, 2, .., n\}$) form an ideal of T. In the quotient ring T/A two elements $\alpha + A = \beta + A$ iff the matrices $\alpha, \beta \in T$ have respectively the same elements on the diagonal. Hence we can choose a complete set of representatives for T/A formed only by diagonal matrices (i.e. which have non-zero coefficients only on the diagonal: $i \neq j \Rightarrow a_{ij} = 0$). The coefficients commute (K is a field) so that the commutativity of T/A is readily checked. Obviously, T is not necessarily commutative.

Ex. 2.21 For the solution some linear algebra results are needed; if R is a ring with identity and $A \in \mathcal{M}_n(R)$ we call *elementary row transformations* the following operations:

(i) the multiplication of any row by any non-zero element, or,

(ii) the addition of any multiple of one row to another row (similarly we define elementary *column* transformations). A simple exercise shows that the interchange of any two rows (or columns) can be realised by a suitable composition of the above operations. The following result is needed:

Theorem.- Any elementary row (column) transformation on a matrix $A \in M_n(R)$ can be realized by a left (right) multiplication of A with a matrix obtained from I_n by the same transformation. $\square$

As consequence of this result we observe that any ideal in $M_n(R)$ is closed under elementary row or column transformations and, as mentioned above, also under interchange of any two rows or columns. In what follows we will denote by

$$E_{ij}^r = \begin{bmatrix} 0 & . & . & 0 & 0 & 0 & . & . & 0 \\ . & . & . & . & . & . & . & . & . \\ 0 & . & . & 0 & 0 & 0 & . & . & 0 \\ 0 & . & . & 0 & r & 0 & . & . & 0 \\ 0 & . & . & 0 & 0 & 0 & . & . & 0 \\ . & . & . & . & . & . & . & . & . \\ 0 & . & . & 0 & 0 & 0 & . & . & 0 \end{bmatrix}$$

the matrix with r on the i-th row and the j-th column and only zero entries elsewhere.

a) Let $\mathcal{A}$ be an arbitrary ideal of $M_n(R)$ and $A = \{a_{11}|(a_{ij}) \in \mathcal{A}\}$. We first prove that A is an ideal of R. The ideal $\mathcal{A}$ being closed under subtraction, $a_{11}, b_{11} \in A$ implies $a_{11} - b_{11} \in A$. Using left and right multiplication with the matrices with at most one non-zero coefficient mentioned above, say E_{11}^r, we deduce that $a_{11} \in A$ implies $r.a_{11}, a_{11}.r \in A$ and hence A is an ideal in $M_n(R)$. In what follows we verify the equality $\mathcal{A} = M_n(A)$. If $(a_{ij}) \in \mathcal{A}$ then by interchanges of 2 rows and 2 columns we obtain another matrix which also belongs to $\mathcal{A}$ (as we have noticed above) and has the element a_{ij} in the left-upper corner. But then $a_{ij} \in A$ and $(a_{ij}) \in M_n(A)$. Conversely, let $(a_{ij}) \in M_n(A)$. The element $a_{ij} \in A$ so that let $V_{ij} \in \mathcal{A}$ a matrix which has a_{ij} as left-upper corner (for every $i, j \in \{1, 2, .., n\}$). Now, the products $E_{11}^1 V_{ij} E_{11}^1$ are matrices with only one non-zero element, namely a_{ij}, on the first row and first column which also belong to $\mathcal{A}$, this being an ideal of $M_n(R)$. Again by interchange of 2 rows and 2 columns we obtain another matrices in $\mathcal{A}$, say W_{ij}, which have the only non-zero element a_{ij} on the i-th row and the j-th column. But obviously $(a_{ij}) = \sum_{i,j} W_{ij} \in \mathcal{A}$.

b) The canonical projection $p_A : R \to R/A$ naturally extends to a surjective ring homomorphism $\bar{p} : M_n(R) \to M_n(R/A)$ (a careful reader remarks that $M_n(R) = \{u|u : I \times I \to R\}$ where $I = \{1, 2, .., n\}$ and $M_n(R/A) = \{\bar{u}|\bar{u} = p_A \circ u\}$). The isomorphism is obtained using a well-known isomorphism theorem and (a) for the kernel of $\bar{p}$.

c) Obviously $2\mathbb{Z}$ is a ring without identity (together with the usual operations in $\mathbb{Z}$). The set $\left\{ \begin{bmatrix} 4x & 2y \\ 2z & 2t \end{bmatrix} \mid x,y,z,t \in \mathbb{Z} \right\}$ is an ideal in $M_2(2\mathbb{Z})$ which differs from $M_2(2n\mathbb{Z}), n \in \mathbb{N}$, $2n\mathbb{Z}$ being the only ideals of $2\mathbb{Z}$, so that (a) fails in rings without identity.

d) No. If A is a left ideal of R one checks that $M_n(A)$ is a left ideal in $M_n(R)$. But the converse does not hold. For instance, it is known that K is a division ring iff $K^2 \neq \{0\}$ and K has no proper one-side ideals. But in $M_n(K)$ the subset $\{(a_{ij}) \in M_n(K) \mid a_{i1} = 0, 1 \leq i \leq n\}$ is a proper left ideal.

Ex. 2.22 One can use the previous exercise: $\mathbb{R}$ being a field, it has only the trivial ideals.

Another solution. Let $E_{11} = \begin{pmatrix} 1 & 0 \\ 0 & 0 \end{pmatrix}, E_{12} = \begin{pmatrix} 0 & 1 \\ 0 & 0 \end{pmatrix}, E_{21} = \begin{pmatrix} 0 & 0 \\ 1 & 0 \end{pmatrix}, E_{22} = \begin{pmatrix} 0 & 0 \\ 0 & 1 \end{pmatrix}$. The following relations hold: $E_{ij} \cdot E_{kl} = \begin{cases} E_{il} & \text{if } j = k \\ 0 & \text{if } j \neq k \end{cases}$

Each matrix $A = \begin{pmatrix} a_{11} & a_{12} \\ a_{21} & a_{22} \end{pmatrix}$ is represented as $A = a_{11}E_{11} + a_{12}E_{12} + a_{21}E_{21} + a_{22}E_{22}$.

Let I be a nonzero ideal of $M_2(\mathbb{R})$. If $0 \neq U = \begin{pmatrix} u_{11} & u_{12} \\ u_{21} & u_{22} \end{pmatrix} \in I$ then the products $E_{ki} \cdot U \cdot E_{jl} = u_{ij}E_{kl} \in I$. At least one coefficient, say u_{12}, is not zero. Then $E_{kl} \cdot U \cdot E_{2l} = u_{l2}E_{kl} \in I$ and hence $a_{kl}E_{kl} = (a_{kl}u_{12}^{-1}E_{kk}) \cdot (u_{l2}E_{kl}) \in I$. Finally $A \in I$ and so $I = M_2(\mathbb{R})$.

Ex. 2.23 Consider the function $\mathcal{P}(X) \to \mathcal{P}(X \setminus Y)$ defined as $f(A) = A \cap (X \setminus Y) = A \setminus Y$ for each $A \in \mathcal{P}(X)$. Clearly, f is surjective but not injective. We verify that f is a (unital, $f(X) = X \setminus Y$ see 4.1) ring homomorphism: $f(A \Delta B) = (A \Delta B) \setminus Y = (A \setminus Y) \Delta (B \setminus Y) = f(A) \Delta f(B)$ and $f(A \cap B) = (A \cap B) \setminus Y = (A \setminus Y) \cap (B \setminus Y) = f(A) \cap f(B)$.

a) By a well-known isomorphism theorem we obtain $\mathcal{P}(X \setminus Y) = im(f) \cong \mathcal{P}(X)/\ker(f) = \mathcal{P}(X)/\mathcal{P}(Y)$. Indeed $\ker(f) = \{A \in \mathcal{P}(X) \mid f(A) = \emptyset\} = \mathcal{P}(Y)$ so that for every proper subset Y of X, $\mathcal{P}(Y)$ is an ideal of $\mathcal{P}(X)$.

b) Let I be an ideal of $\mathcal{P}(X)$. First check $A \cup B = (A\Delta B)\Delta(A\cap B)$ and hence, for $A, B \in I$ one has also $A \cup B \in I$. By induction, I is also closed to finite unions and moreover I is closed to lower bounds. Indeed, if $C \leq A \in I$ then $C = C \cap A \in I$. Now, let $Y = \bigcup_{A \in I} A$ element of I as finite union (X is finite). Finally we verify the equality $I = \mathcal{P}(Y)$. If $A \in I$ then $A \subseteq Y$ and hence $A \in \mathcal{P}(Y)$. Conversely, if $B \in \mathcal{P}(Y)$ then $B \leq Y \in I$ and hence (as we saw above) $B \in I$.

c) Let X be infinite and $I = \{A \in \mathcal{P}(X) | A \text{ finite}\}$; in order to verify that I is an ideal first observe that in the additive group of the ring $(\mathcal{P}(X), \Delta)$ we have $-B = B$ so that $A - B = A\Delta B \subseteq A \cup B$ is also finite and hence $A - B \in I$. Further, if $A \in I$ and $C \in \mathcal{P}(X)$ then $A \cap C \subseteq A$ is also finite so that $A \cap C \in I$. Finally, if $I = \mathcal{P}(Y)$ would hold for a suitable $Y \subseteq X$, from $Y \in \mathcal{P}(Y)$ we see that Y is finite. X being infinite for $x_0 \in X\backslash Y$ obviously $Y \cup \{x_0\}$ is also finite so that $Y \cup \{x_0\} \in I$ and hence $Y \cup \{x_0\} \in \mathcal{P}(Y)$, a contradiction.

Ex. 2.24 Considering the canonical ring homomorphisms $p_I : R \to R/I$, $p_J : R \to R/J$, the so called property of universality of the direct product induces $\alpha : R \to R/I \times R/J$ a ring homomorphism which has the kernel $I \cap J$. Using a Noether isomorphism theorem one deduces that the canonical ring homomorphism $R/I \cap J \to R/I \times R/J$ is an isomorphism iff α is surjective. If I and J are comaximal we prove that α is surjective. For an arbitrary element $(a + I, b + J) \in R/I \times R/J$ we consider $a - b \in R = I + J$. If $i \in I$ and $j \in J$ are such that $a - b = -i + j$ denote by $r = a + i = b + j$. We immediately get $\alpha(r + (I \cap J)) = (a + I, b + J)$ so α is surjective. Conversely, the surjectivity of α is equivalent to: $\forall a, b \in R, \exists r \in R : a - r \in I, b - r \in J$. Taking $b = 0$ we obtain $a = (a - r) + r \in I + J$, that is $R = I + J$.

Ex. 2.25 The ideal I being finitely generated, let $I = Ra_1 + .. + Ra_n$. Surely also $I = Ia_1 + .. + Ia_n$ holds (indeed, one has to verify the two inclusions using $IR = I$ and $I^2 = I$) and then for each $k, 1 \leq k \leq n$ there is an element $b_k \in I$ such that $(1 - b_k)I \subseteq Ia_k + .. + Ia_n$. One verifies this last assertion by induction on k (as for $k = 1$, the element $b_1 = 0$ is obviously suitable; moreover, if an element $b_k \in I$ is chosen there is an element (prove this !) $b_{k+1} \in I$ such that $(1 - b_{k+1})I \subseteq Ia_{k+1} + .. + Ia_n$). Finally , $e = b_{n+1} \in I$ is the required idempotent.

Indeed, $(1 - b_{n+1})I \subseteq 0$ implies $(1 - e)e = 0$ or $e^2 = e$ and $I = Ie$ and hence $I = Re$.

Ex. 2.26 It is sufficient to verify that each ideal generated by 2 elements, say $a, b \in R$, is principal. Using the solution of 17.14 we can suppose that these elements are idempotent and $a \notin Rb, b \notin Ra$. Moreover, Ra is a direct summand in R and so $b = b' + b''$, where $b' \in Ra$ and $ab'' = 0$. Then $Rb = Rb' \oplus Rb''$ because $b', b'' \in Rb$. We surely have $Ra + Rb = Ra + Rb''$ and this is precisely $R(a + b'')$ (i.e. principal) because $a = a(a + b'')$ and $b'' = b''(a + b'')$ (one checks that b' and b'' are also idempotents).

Ex. 2.27 (a) Obvious.

(b) The fractions being irreducible, for each $x = \frac{m}{n} \in \mathbb{Q}$ at least one of m or n is relatively prime with p and so $x \in \mathbb{Q}^{(p)}$ or $x^{-1} \in \mathbb{Q}^{(p)}$.

(c) Let S be a subring of $\mathbb{Q}$ so that $\mathbb{Q}^{(p)} \subset S$ (strict inclusion). We show that $\mathbb{Q} \subseteq S$ (and hence $\mathbb{Q} = S$). Let $r = \frac{m}{n} \in \mathbb{Q}$. If $(n; p) = 1$ then $r \in \mathbb{Q} \subseteq S$. If $(n; p) \neq 1$ then let $n = p^s t$ with $(t; p) = 1$. By our hypothesis let $r' = \frac{m'}{n'} \in S \backslash \mathbb{Q}^{(p)}$ where $(n'; p) \neq 1$ and hence $(m'; p) = 1$. Then $\frac{1}{m'} \in \mathbb{Q}^{(p)} \subseteq S$ and $\frac{1}{m'}.r' = \frac{1}{n'} \in S$. Now if $n' = p^v u$ with $(u; p) = 1$ we also have $\frac{1}{p^v} \in S$ and hence $\frac{1}{p^k} \in S$ with $k \in \mathbb{N}^*, k \leq v$. But hence, S being a subring, also $\frac{1}{p^k} \in S$ for $k \geq v$ and so $\frac{1}{p^s} \in S$. Finally, $r = \frac{m}{t}.\frac{1}{p^s} \in S$ because $\frac{m}{t} \in \mathbb{Q}^{(p)} \subseteq S$.

(d) Let $I \neq 0$ be an ideal of $\mathbb{Q}^{(p)}$. If I contains a rational number $\frac{m}{n}$ with $(m; p) = 1$ then $\frac{m}{n}$ is a unit and $I = \mathbb{Q}^{(p)}$. In the remaining case, let p^n be the greatest power of p which divides all the numerators of the elements from I. Then $I \subseteq p^n.\mathbb{Q}^{(p)}$. Conversely, let $r = p^n.\frac{s}{t} \in I$ such that $(s; p) = 1$. Then $\frac{t}{s} \in \mathbb{Q}^{(p)}$ and $p^n = \left(\frac{t}{s}.p^n\right).\frac{s}{t} \in I$ and hence $p^n\mathbb{Q} \subseteq I$.

(e) For every $p \in \mathbb{P}$ we clearly have $\mathbb{Z} \subseteq \mathbb{Q}^{(p)}$. Conversely, if $\frac{m}{n} \in \bigcap_{p \in \mathbb{P}} \mathbb{Q}^{(p)}$ then each $p \in \mathbb{P}$ divides n. Hence $n \in \{\pm 1\}$ and $\frac{m}{n} \in \mathbb{Z}$.

Chapter 3

Zero Divisors

Ex. 3.1 If $e \in R$ is an idempotent element such that $e \notin \{0,1\}$ then obviously $e.(1-e) = 0$.

Ex. 3.2 Denote by $X = \{x \in [0,1] | g(x) = 0\}$ for $0 \neq g \in R$ such that $fg = 0$. This is a closed set (g being continuous) and hence f vanishes on the complement of X in $[0,1]$ and then also on an open interval from $[0,1]$. Conversely, if f vanishes on the open interval (a,b) then the function $g \in R$ defined by

$$g(x) = \begin{cases} (x-a)(x-b) & \text{for } x \in (a,b) \\ 0 & \text{otherwise} \end{cases}$$

is a zero divisor for f.

Let $f \in R$ be an idempotent element. Hence $f^2 = f$ or $f.(1-f) = 0$ denoting by 0 resp. 1 the corresponding constant functions. Then $f(x) = 0$ or $f(x) = 1$ for each $x \in [0,1]$. But f being continuous only $f = 0$ and $f = 1$ are possible. Similarly, the unique nilpotent element is 0.

Ex. 3.3 (i) If A, B are finite subsets in M clearly $A + B = (A - B) \cup (B - A)$ and $A \cap B$ are also finite.

(ii) For each finite subset A from M if $x \in M - A$ (which is nonvoid, M being infinite) then surely $A \cap \{x\} = \emptyset$ and so A is a zero divisor in $\mathcal{P}_0(M)$;

(iii) If $M \neq N \in \mathcal{P}(M)$ and $x \in M - N$ then $N \cap \{x\} = \emptyset$ and hence N is a zero divisor in $\mathcal{P}(M)$.

Ex. 3.4 If $\overline{mx}, \overline{my} \in H$ then $\overline{mx} \cdot \overline{my} = \overline{m \cdot mxy}$ so that from $(m; n) = 1, (x, n) = 1$ and $(y; n) = 1$ we have $(mxy; n) = 1$ or $\overline{m \cdot mxy} \in T$. Hence T is closed under multiplication. Further, let $t_{\overline{a}} : T \to T, t_{\overline{a}}(\overline{x}) = \overline{a}.\overline{x}$, the (left) translation with $\overline{a} \in T$. This map is injective: if $\overline{x}, \overline{y} \in T$ and $t_{\overline{a}}(\overline{x}) = t_{\overline{a}}(\overline{y})$ then $\overline{a}(\overline{x} - \overline{y}) = \overline{0}$ and if $\overline{a} = \overline{ma_1}, \overline{x} = \overline{mx_1}, \overline{y} = \overline{my_1}$ then $(a_1; n) = (x_1; n) = (y_1; n) = 1$. Since $q = mn$ divides $m^2 a_1 (x_1 - y_1)$ we deduce that n divides $ma_1 (x_1 - y_1)$ and hence n divides $x_1 - y_1$ (because n is relatively prime with m and a_1). Hence $q = mn$ divides $mx_1 - my_1$ or $\overline{x} = \overline{mx_1} = \overline{my_1} = \overline{y}$. T being finite $t_{\overline{a}}$ is also surjective and then one checks immediately that $(T, \cdot)$ is a group. Finally, the elements from T are zero divisors in $\mathbb{Z}_q$: for each $\overline{mx} \in T$ one has $\overline{mx} \cdot \overline{n} = \overline{0}$.

Remark. One verifies that $(T, \cdot) \cong U(\mathbb{Z}_n)$ the multiplicative group of units.

Ex. 3.5 Let $l \in R$ (resp. $r \in R$) be a left (resp. right) non-zero divisor. Consider the corresponding traslations $t_l : R \to R, t_l(x) = lx$ resp. $t_r'(x) = xr, \forall x \in R$. These are injective (indeed, e.g. $t_l(x) = t_l(y) \Rightarrow l.(x - y) = 0 \Rightarrow x - y = 0 \Rightarrow x = y$) functions and, R being finite, also surjective functions. Hence, there are elements $a, b \in R$ such that $t_l(a) = l$ and resp. $t_r'(b) = r$. Then $t_l(ax) = lax = lx = t_l(x)$ and $t_r'(xb) = xbr = xr = t'(x)$ so that again by injectivity $ax = x = xb, \forall x \in R$. So $a = ab = b$ and the corresponding element is an identity for R.

Ex. 3.6 If $s = ar, s' = a'r'$ then $ss' = aa'rr'$ and so $aa' \in T$, a multiplicative system. Further, if $a \in T$ would be a zero divisor then for a suitable $r \in R$ the corresponding element $s = ar$ would be a zero divisor too.

Ex. 3.7 For $r = 0$ nothing is to be proved. If $r \neq 0$ consider an arbitrary element $x \in I$. From $(xr)a = 0$ follows $xr = 0$ (because $a \neq 0$ and I has no non-zero divisors) and then $x(rx) = 0$. Again, R having no non-zero divisors, $rx = 0$ for $x \neq 0$ ($x = 0$ is evident).

Ex. 3.8 The intersection $J = l(I) \cap r(I)$ is an ideal (see 2.3) in R, so we can consider the quotient ring. If $(r + J)(s + J) = J$ are zero divisors in R/J then $rs \in J$ and then $(rs)a = a(rs) = 0, \forall a \in I$. Now, if $sa \neq 0$ then one uses the previous exercise for $rI = Ir = (0)$, i.e. $r \in J$. Finally, if $sa = 0$ one verifies in an analogous way that $s \in J$; hence $r + J = J$ or $s + J = J$ and R/J has no non-zero divisors.

Ex. 3.9 With the notations of the previous exercise, $(I + J)/J \cong I$ because clearly $I \cap J = (0)$ and one uses a Noether isomorphism theorem. Finally, apply the previous exercise to the Dorroh extension and remark that if R has no non-zero divisors then $J = l(I) = r(I) = (0)$.

Ex. 3.10 If $f = a_0 + a_1 X + .. + a_n X^n$ is a zero divisor let $g = b_0 + b_1 X + .. + b_m X^m$. polynomial of lowest degree with $g \neq 0$ and $f \cdot g = 0$. Then $a_n b_m = 0$ and surely $a_n g = a_n b_0 + .. + a_n b_{m-1} X^{m-1}$ and $(a_n g) f = 0$ with $\deg(a_n g) \leq m - 1 < m$. Using the minimality of g we must have $a_n g = 0$. Next, consider $f_1 = a_0 + a_1 X + .. + a_{n-1} X^{n-1}$. Clearly $fg = 0$ and $a_n g = 0$ imply $f_1 g = 0$ and then $a_{n-1} b_{m-1} = 0$. As before $a_{n-1} g = 0$ and hence, step by step, $a_k g = 0$ for each $0 \leq k \leq n$. Finally, $a_k b_s = 0$ for each $0 \leq s \leq m$ and so $b_m \cdot f = 0$. The converse is obvious.

Ex. 3.11 Let $a \in R$ be a left non-zero divisor. The ring R being finite the injective map $t_a : R \to R, t_a(x) = ax, \forall x \in R (ax = ay \Rightarrow x = y$ because a is a non-zero divisor$)$ is also surjective. Hence there is an element $b \in R$ such that $t_a(b) = 1$ or $ab = 1$. The element b cannot be a left zero divisor because $bc = 0$ (with $c \neq 0$) would imply $c = 1.c = (ab)c = a(bc) = a.0 = 0$. As above t_b is surjective and hence there is an element $d \in R$ such that $t_b(d) = 1$ or $bd = 1$. But $a = a.1 = a(bc) = (ab)c = 1.c = c$ and hence $ba = 1$ so that a is a unit.

Ex. 3.12 It suffices to show that, equivalently, if $A = \begin{pmatrix} a & b \\ c & d \end{pmatrix}$ is not a unit in $\mathcal{M}_2(K)$ then A is a left zero divisor. We can suppose that at least one coefficient is nonzero, say $a \neq 0$. First, if $c = d = 0$ then simply $\begin{pmatrix} a & b \\ c & d \end{pmatrix} \begin{pmatrix} b & 0 \\ -a & 0 \end{pmatrix} = \begin{pmatrix} 0 & 0 \\ 0 & 0 \end{pmatrix}$. Finally, if c, d are not both zero, suppose that, say $c \neq 0$. Then we have $\begin{pmatrix} a & b \\ c & d \end{pmatrix} \begin{pmatrix} -a^{-1}b & -c^{-1}d \\ 1 & 1 \end{pmatrix} = \begin{pmatrix} 0 & -ac^{-1}d + b \\ -ca^{-1}b + d & 0 \end{pmatrix} = \begin{pmatrix} 0 & 0 \\ 0 & 0 \end{pmatrix}$ because $ad - bc = \det(A) =$

0, by hypothesis. Finally, if K is not a field the assertion remains not valid: take $2 \cdot I_2 \in \mathcal{M}_2(\mathbb{Z})$.

Ex. 3.13 $A = \begin{pmatrix} a + bi & c + di \\ -c + di & a - bi \end{pmatrix} = aI_2 + bI + cJ + dK$ where $I = \begin{pmatrix} i & 0 \\ 0 & -i \end{pmatrix}, J = \begin{pmatrix} 0 & 1 \\ -1 & 0 \end{pmatrix}, K = \begin{pmatrix} 0 & i \\ i & 0 \end{pmatrix}$ together with $I^2 = J^2 = K^2 = -I_2, IJ = -JI = K, JK = -KJ = I, KI = -IK = J$, are representations of matrices which simplify the verifications needed in order to prove that, together with the usual laws of addition and multiplication, the set M is a ring. If $A \in M, A \neq 0$ then $\det(A) = a^2 + b^2 + c^2 + d^2 \neq 0$ and so A has an inverse (which also belongs to M). This ring is then a division ring, obviously without zero divisors.

Remark. See also 7.12,**3**.

Ex. 3.14 Indeed, if $ba \neq 0$ then a is left zero divisor and if $ba = 0$ then a is right zero divisor.

Ex. 3.15 If $a \in R$ is not a zero divisor, from $aba = a$ for a suitable $b \in R$ we have $(ab - 1)a = 0$ and so $ab = 1$.

Ex. 3.16 No. Generally they do not even form a subgroup of the additive group $(R, +)$. For instance, in $\mathbb{Z}_6$ the classes $\bar{2}$ and $\bar{3}$ are zero divisors but $\bar{3} - \bar{2} = \bar{1}$ is not.

Remark. A ring is called *subdirectly irreducible* if the intersection of all the nonzero ideals is not zero. One can prove that in a subdirectly irreducible ring the answer is affirmative.

Ex. 3.17 We use 15.6: the multiplicative system S_0 of all the non-zero divisors is saturated (indeed, if xy is non-zero divisor obviously x and y have the same property) and so $D = \cup\{P|P \text{ prime ideal }, P \cap S_0 = \emptyset\}$. Moreover, we can show that D contains all the minimal prime ideals. Indeed, using 13.4, any maximal multiplicative system S contains S_0 (otherwise, for an element $a \in S_0, a \notin S$, the multiplicative system $\{1, a, a^2, .., a^n, ..\} \cdot S \supset S$ and S would not be maximal). Finally, for a minimal prime ideal $P, R\backslash P$ is a maximal multiplicative system and from $S_0 \subseteq R\backslash P$ we infer $P \subseteq D$.

Ex. 3.18 For $f \in D\langle X\rangle, f = \sum_{n=0}^{\infty} a_n X^n$ we call the *order* $o(f)$ the lowest $n \in \mathbb{N} : a_n \neq 0$ and $o(0) = -\infty$. One easily checks that $o(f+g) \geq \min(o(f), o(g))$ and $o(fg) \geq o(f) + o(g)$. If $g = \sum_{n=0}^{\infty} b_n X^n$ and D is an integral domain, $a_{o(f)} b_{o(g)}$ is the only non-zero product $a_s b_t$ such that $s + t = o(f) + o(g)$. Hence if $fg = \sum_{n=0}^{\infty} c_n X^n$ then $c_{o(f)+o(g)} = a_{o(f)} b_{o(g)} \neq 0$ and $o(f) + o(g) = o(fg)$. In particular, $o(fg) = -\infty$ iff at least one from $o(f), o(g)$ is $-\infty$, i.e. $fg = 0$ iff $f = 0$ or $g = 0$.

Ex. 3.19 Neither implication holds. For $p_4 : \mathbb{Z} \to \mathbb{Z}_4, p_4(x) = [x]_4$, we have $p_4(2) = [2]_4$ where $[2]_4 \in \mathbb{Z}_4$ is a zero divisor ($[2]_4 \cdot [2]_4 = [0]_4$) but $2 \in \mathbb{Z}$ is not. Conversely, let $f : \mathbb{Z}_6 \to \mathbb{Z}_3, f([x]_6) = [x]_3$,(see 4.9, the translation with $[1]_3$). Obviously, $f([2]_6) = [2]_3$ and $[2]_6$ is zero divisor in $\mathbb{Z}_6$ and $[2]_3$ in $\mathbb{Z}_3$ is not (this is a field and $[2]_3 \neq [0]_3$).

Chapter 4

Ring Homomorphisms

Ex. 4.1 If a' is arbitrary in $f(R)$ there is an element $a \in R$ such that $a' = f(a)$. Then $a'.f(1) = f(a).f(1) = f(a.1) = f(a) = a'$ and similarly $f(1).a' = a'$ so that $f(1)$ is the identity in $f(R)$. An easy (but not trivial!) counterexample is $f : \mathbb{Z}_{12} \to \mathbb{Z}_{12}, f(\overline{x}) = \overline{4x}, \forall \overline{x} \in \mathbb{Z}_{12}$. Here $\overline{4}$ is the identity in $f(\mathbb{Z}_{12}) = \{\overline{0}, \overline{4}, \overline{8}\}$ and surely $\overline{1} \neq \overline{4}$.

As for the application, $f(R) = R'$ and by uniqueness $f(1) = 1'$.

Ex. 4.2 Indeed, ker f being ideal in the simple ring K, we have ker $f = K$ or ker $f = \{0\}$. Hence f is either trivial or injective.

Ex. 4.3 $\mathbb{Q}$ being a field (and hence simple), the ring homomorphisms f, g must be either trivial or injective (see the previous exercise). If $f = g = 0$ there is nothing to prove. From $f(n) = g(n), \forall n \in \mathbb{Z}$ we deduce that the only case left is f, g both injective. In this case the subrings $f(\mathbb{Q}), g(\mathbb{Q})$ are fields in R with the same identity $1' = f(1) = g(1)$. Now if A is the subring of R generated by $f(\mathbb{Q}) \cup g(\mathbb{Q})$ then $1'$ is also identity in A and f, g considered as homomorphisms $\mathbb{Q} \to A$ are unital. Hence $f(n)$ and $g(n)$ are units in A for each $n \in \mathbb{Z}^*$ and $[f(n)]^{-1} = f\left(\frac{1}{n}\right), [g(n)]^{-1} = g\left(\frac{1}{n}\right)$. But then from $f(n) = g(n)$ we have also $[f(n)]^{-1} = [g(n)]^{-1}$ and hence $f\left(\frac{1}{n}\right) = g\left(\frac{1}{n}\right)$. Finally, one gets $f\left(\frac{m}{n}\right) = mf\left(\frac{1}{n}\right) = mg\left(\frac{1}{n}\right) = g\left(\frac{m}{n}\right), \forall \frac{m}{n} \in \mathbb{Q}$.

Ex. 4.4 Obviously $(3; 5) = 1$ and $2.3 - 1.5 = 1$. If $([a]_3, [b]_5) \in \mathbb{Z}_3 \times \mathbb{Z}_5$ then we consider $n = 2.3b - 1.5a$. Hence $x - a = 3.2(b-a), x - b = 5.1(a-$

b) or $3|x - a, 5|x - b$. Then $[x]_3 = [a]_3$, $[x]_5 = [b]_5$ and $f(x) = ([a]_3, [b]_5)$. Moreover, $\ker f = 3\mathbb{Z} \cap 5\mathbb{Z} = [3; 5]\,\mathbb{Z} = 15\mathbb{Z}$.

Generalization. If $(n; m) = 1$ and $un + vm = 1$ then $x = unb + vma$ does the same job for the surjectivity of f and it only remains to verify that $\ker f = nm\mathbb{Z}$.

Ex. 4.5 The composition is surjective iff for every $y \in \mathbb{Z}\left[\sqrt{d}\right]$ there are elements $u \in \mathbb{Z}, v \in \mathbb{Z}\left[\sqrt{d}\right]$ such that $y = u + xv$. If $x = m + n\sqrt{d}, y = a + b\sqrt{d}$ and $v = s + t\sqrt{d}$ then our composition is surjective iff for each $b \in \mathbb{Z}$ there are $s, t \in \mathbb{Z}$ such that $b = ms + nt$. But this last condition is equivalent with $(m; n) = 1$. As for the special case, the kernel of the composition being $(x) \cap \mathbb{Z}$, using an isomorphism theorem, the canonical homomorphism is isomorphism iff $(m; n) = 1$.

Ex. 4.6 Take $i : \mathbb{Z} \to \mathbb{Q}, i(n) = n, \forall n \in \mathbb{Z}$, the inclusion ring homomorphism. Obviously, $\mathbb{Z}$ is an (improper) ideal of $\mathbb{Z}$ but $i(\mathbb{Z}) = \mathbb{Z}$ is no ideal in $\mathbb{Q}$.

Remark. Surjective ring homomorphisms preserve also the ideals (and not only subrings as it is well-known in general).

Ex. 4.7 More generally, for each ring with identity R there a unique unital ring homomorphism $f : \mathbb{Z} \to R$ i.e. $f(n) = n.1, \forall n \in \mathbb{Z}$. Hence only $1_{\mathbb{Z}}$ is a unital ring endomorphism. In order to determine the ring homomorphisms $\mathbb{Z} \to \mathbb{Q}$ we recall that the group homomorphisms $f : \mathbb{Z} \to G$ where G is an abelian group, are determined by $f(1) \in G$ i.e. $f(n) = n.f(1)$, so for such group homomorphisms with $f(1) \in \mathbb{Q}$ one has to find which are also ring homomorphisms or, equivalently, which are the idempotent elements of $\mathbb{Q}$ (f ring homomorphism$\Leftrightarrow f(1)$ idempotent). But the only idempotent elements in a ring with identity without zero divisors are 0 and 1.

Hence the only ring homomorphisms are the zero and the identity homomorphisms.

Ex. 4.8 We prove that all the ring endomorphisms of $\mathbb{Z}_n$ are the translations with the idempotent elements of $\mathbb{Z}_n$ (and one uses 1.22). First, if $f : \mathbb{Z}_n \to \mathbb{Z}_n$ is a ring endomorphism then it is also a group endomorphism and hence $f(\overline{m}) = f(m.\overline{1}) = mf(\overline{1}) = \overline{m}.f(\overline{1}) = t_{f(\overline{1})}(\overline{m})$ so that $f = t_{f(\bar{1})}$. But $f(\bar{1}) \in Id(\mathbb{Z}_n)$ because $f(\bar{1}).f(\bar{1}) = f(\bar{1}.\bar{1}) = f(\bar{1})$. Conversely, if $\bar{a} \in Id(\mathbb{Z}_n)$ then $t_{\bar{a}}(\bar{x} + \bar{y}) = t_{\bar{a}}(\bar{x}) + t_{\bar{a}}(\bar{y})$ and $t_{\bar{a}}(\bar{x}.\bar{y}) = t_{\bar{a}}(\bar{x}).t_{\bar{a}}(\bar{y})$ hold.

Ex. 4.9 As above one can verify that the required ring homomorphisms are the translations with the idempotent elements $[a]_m \in \mathbb{Z}_m$ such that $n.[a]_m = [0]_m$ holds in $\mathbb{Z}_m$. Such a translation is now $t_{[a]_m}$: $\mathbb{Z}_n \to \mathbb{Z}_m$ defined by $t_{[a]_m}([x]_n) = [a.x]_m$. The only non obvious thing to be verified is that this translation is well-defined: $[x]_n = [y]_n$ and $n.[a]_m = [0]_m \Rightarrow [a.x]_m = [a.y]_m$. Indeed, from $x - y = nt$ and $na = ms$ we obtain at once $a.x - a.y = mst$.

Ex. 4.10 The affirmative way follows by simple computations. None of the four converses is true. Counterexamples: we denote by $p_n : \mathbb{Z} \to \mathbb{Z}_n$ the projection $p_n(x) = \bar{x}$, $\forall x \in \mathbb{Z}$. For $n = 6$, 3 is not idempotent in $\mathbb{Z}$ but $\bar{3}$ is idempotent in $\mathbb{Z}_6$; for $n = 4$, 2 is not nilpotent in $\mathbb{Z}$ but $\bar{2}$ is nilpotent in $\mathbb{Z}_4$; for $n = 5$ every integer $a \notin \{-1, 0, 1\}$ is not a unit in $\mathbb{Z}$ but $\bar{a}$ is a unit in $\mathbb{Z}_5$ (this being a field). For the last example it suffices to give an example of non-commutative ring R with a proper commutative quotient ring R/I and to take $p_I : R \to R/I$ the corresponding projection. But this is 2.22.

Ex. 4.11 (a) Indeed, $f(x)^2 = f(x^2) = f(1+1) = f(1) + f(1) = 1' + 1'$ (note that ring isomorphisms are unital 4.1).

(b) $\mathbb{Z}[\sqrt{2}]$ has elements which satisfy the relation from (a) (e.g. $\sqrt{2}$) but $\mathbb{Z}[\sqrt{3}]$ has not. Indeed, $(a + b\sqrt{3})^2 = 1 + 1$ or the system $a^2 + 3b^2 = 2, 2ab = 0$ have no solutions in $\mathbb{Z}$, and one applies (a).

Generalization. If $d \neq e$ are square-free positive integers the same holds for $\mathbb{Z}[\sqrt{d}]$ and $\mathbb{Z}[\sqrt{e}]$.

Ex. 4.12 Obviously R is a subring: $A, B \in R \Rightarrow A - B, A.B \in R$; since

$$\begin{pmatrix} 0 & 1 \\ 0 & 0 \end{pmatrix}\begin{pmatrix} 0 & 0 \\ 0 & 1 \end{pmatrix} = \begin{pmatrix} 0 & 1 \\ 0 & 0 \end{pmatrix} \neq \begin{pmatrix} 0 & 0 \\ 0 & 0 \end{pmatrix} = \begin{pmatrix} 0 & 0 \\ 0 & 1 \end{pmatrix}\begin{pmatrix} 0 & 1 \\ 0 & 0 \end{pmatrix} \quad R$$

has the required properties. Next, $\begin{pmatrix} 0 & a \\ 0 & b \end{pmatrix}^2 = \begin{pmatrix} 0 & ab \\ 0 & b^2 \end{pmatrix} = 0 \Leftrightarrow$

$b = 0$ so that $I = \left\{ \begin{pmatrix} 0 & a \\ 0 & 0 \end{pmatrix} \mid a \in \mathbb{Q} \right\}$. Now, $A, B \in I, C \in R \Rightarrow$

$A-B, CA, AC \in I$ are easily verified. Finally $\begin{pmatrix} 0 & a_1 \\ 0 & b_1 \end{pmatrix} - \begin{pmatrix} 0 & a_2 \\ 0 & b_2 \end{pmatrix} \in$

$I \Leftrightarrow b_1 = b_2$ so that the map $f : R \to \mathbb{Q}, f\left(\begin{pmatrix} 0 & a \\ 0 & b \end{pmatrix}\right) = b, \forall b \in \mathbb{Q}$

is a surjective ring homomorphism with $\ker f = I$; so the required isomorphism is given by a well-known isomorphism theorem.

Ex. 4.13 f being a ring homomorphism $f(A + \ker(f)) = f(A)$ so that $A + \ker(f) \subseteq f^{-1}(f(A))$. Conversely, if $r \in f^{-1}(f(A))$ then $f(r) = f(a)$ for a suitable $a \in A$ and $f(r-a) = 0$ or $r - a \in \ker(f)$. Hence $r \in a + \ker(f) \subseteq A + \ker(f)$. The following equivalences prove the

remaining equality $r \in f^{-1}\left(\bigcap_{i \in I} B_i\right) \Leftrightarrow f(r) \in \bigcap_{i \in I} B_i \Leftrightarrow \forall i \in I : f(r) \in$

$B_i \Leftrightarrow \forall i \in I : r \in f^{-1}(B_i) \Leftrightarrow$
$\quad r \in \bigcap_{i \in I} f^{-1}(B_i).$

Ex. 4.14 Consider the inclusion $f : \mathbb{Z} \to \mathbb{C}[X, Y]$ and M the maximal ideal (see 13.14) of all the polynomials with zero constant term. $f^{-1}(M) = \{0\} \neq \mathbb{Z}$ is not a maximal ideal in $\mathbb{Z}$.

   Remark. If $f : R \to R'$ is a surjective ring homomorphism and M is a maximal ideal in R' then $f^{-1}(M)$ is a maximal ideal in R.

Ex. 4.15 Suppose that such a ring R does exist. R being a subring with identity of $R[X]$, it must be isomorphic with a subring with identity of $(\mathbb{Z}, +, \cdot)$. In particular, $(R, +)$ must be group-isomorphic with a nonzero subgroup of $(\mathbb{Z}, +)$ and hence $\exists n \in \mathbb{N}^*$ such that $R \cong n\mathbb{Z}$. But $1 \in R$ implies $n = 1$ and so $\mathbb{Z} \cong \mathbb{Z}[X]$ a contradiction (e.g. $\mathbb{Z}$ is cyclic and obviously $\mathbb{Z}[X]$ isn't cyclic). Hence there are no such rings.

Chapter 5

Characteristics

Ex. 5.1 If $a \in R$, a Boole ring, $a + a = (a+a)^2 = a+a+a+a$ so that $a + a. = 0$. Hence the required characteristic is 2.

Ex. 5.2 From $char(R) = ord_{(R,+)}(1)$ follows
$$char\,(\mathbb{Z} \times \mathbb{Z}) = char\,(\mathbb{Z} \times \mathbb{Z}_3)$$
$$= char\,(End(\mathbb{Z})) = \infty \text{ and } char\,(End(\mathbb{Z}_3)) = 3.$$

Ex. 5.3 As above $char\,(\mathbb{Z}_m \times \mathbb{Z}_n) = ord([1]_m, [1]_n) = [m;n]$ the lower common multiple (indeed, one uses the following group theoretic result: let x, y be elements in a group G, $xy = yx$, $ord(x) = m$, $ord(y) = n$. If $\langle x \rangle \cap \langle y \rangle = \{1\}$ then $ord(xy) = [m;n]$).
    Generalization. If $char(R) = m$ and $char(R') = n$ one proves in a similar way that $char\,(R \times R') = [m;n]$.

Ex. 5.4 As in the previous exercise $char\,(\mathbb{Q} \times \mathbb{Z}_8) = ord\,(1, [1]_8) = \infty$.

Ex. 5.5 Let K be a division ring. If $char(K) = p$ define $\overline{m} \cdot x = mx, \forall \overline{m} \in \mathbb{Z}_p, \forall x \in K$. The definition does not depend from representatives: $n \in \overline{m} \Rightarrow nx = mx$ and so, simple verifications show that K actually is a vector space over $\mathbb{Z}_p$. If $char(K) = 0$ define $\left(\frac{m}{n}\right)x = m \cdot (nx)^{-1}, \forall \frac{m}{n} \in \mathbb{Q}, \forall x \in K$. This case is similar: one shows that K actually is a vector space over $\mathbb{Q}$.

Ex. 5.6 If $f : \mathbb{Z} \to R$ is a unital ring homomorphism then surely $f(m) = m.1, \forall m \in \mathbb{Z}$. This proves the uniqueness; the existence follows obviously (for a map defined by this formula) by straightforward ring computations. If $char(R) = 0$ then $f(m) = f(l) \Rightarrow m.1 = l.1 \Rightarrow (m - l).1 = 0 \Rightarrow m - l = 0 \Rightarrow m = l$. Finally, if $char(R) = n$ then $m.1 = 0 \Leftrightarrow \exists k \in \mathbb{Z} : m = kn$ and hence $\ker(f) = (n) = n\mathbb{Z}$.

Ex. 5.7 The kernel $\ker(f)$ being an ideal in the division ring K one has $\ker(f) = (0)$ because otherwise $f = 0$ would not be unital. Apply the previous exercise (and especially the uniqueness).

Ex. 5.8 The condition is clearly necessary. Conversely, we have to show that $f(x.y) = f(x).f(y), \forall x, y \in K$. First, we prove that $f(x^2) = f(x)^2, \forall x \in K$. Indeed, this follows from $f((1 + x)^3) = f(1 + x)^3$ using $f(1) = 1'$, $f(x^3) = f(x)^3$ and $char(K') \neq 3$. Finally, from $2xy = (x + y)^2 - x^2 - y^2$, applying f (which is a group homomorphism for the additive subjacent groups) we get $2f(xy) = f((x+y)^2) - f(x^2) - f(y^2) = f(x + y)^2 - f(x)^2 - f(y)^2 =$

$$(f(x) + f(y))^2 - f(x)^2 - f(y)^2 = 2f(x)f(y) \text{ and hence } f(xy) = f(x)f(y) \text{ because } char(K') \neq 2.$$

Ex. 5.9 $char(\mathbb{Z}_p \times \mathbb{Z}_p) = p$ and this is no field (one finds easily zero divisors).

Ex. 5.10 If one uses the representation $0, 1, x, 1 + x, x^2, 1 + x^2, x + x^2, 1 + x + x^2$ together with $2x = 0$, for each $a \in K_8$ one checks $2a = 0$.

Ex. 5.11 The elements of this field (see also 7.3) may be represented as $0, 1, 2, y, 2y, 1 + y, 1 + 2y, 2 + y, 2 + 2y$ where $3y = 0$. For each such element a one verifies $a + a + a = 0$. So the characteristics is 3.

Ex. 5.12 Let $\mathbb{Z}_p[X]$ the corresponding polynomial ring. This is an integrity domain (together with $\mathbb{Z}_p$) and $char(\mathbb{Z}_p[X]) = p$. Let K be the ring of quotients of $\mathbb{Z}_p[X]$. This is obviously an infinite field and one has $p \cdot \overline{(y, y)} = \overline{(py^p, y^p)} = \overline{(0, y^p)}$ the zero element in the ring of quotients. Hence $char(K) = p$. More generally, the ring of quotients of an integrity domain of characteristic p has the same characteristic.

Ex. 5.13 Let $A = (a_{ij})_{1 \leq i,j \leq n}$ and $B = (b_{ij})_{1 \leq i,j \leq n}$. The sum of the elements (also called *the trace*) of the diagonal of the product $A \cdot B$ is $\sum_{i=1}^{n} \sum_{k=1}^{n} a_{ik} b_{ki}$. A simple computation shows that the trace of the product $B \cdot A$ is the same. Hence the trace of $A \cdot B - B \cdot A$ is zero (the "trace" is an additive map). But the trace of I_n is obviously $n \cdot 1 \neq 0$ if the characteristic is 0.

Ex. 5.14 Consider $f : K \to K, f(x) = x^{p^n}, \forall x \in K$. Using $char(K) = p$ one verifies similarly to 8.11 that f is a unital ring homomorphism (between fields) and then f is injective (indeed, $f \neq 0 \Rightarrow \ker(f) \neq K \Rightarrow \ker(f) = (0)$). Hence the given polynomial has at most one root in K.

Ex. 5.15 It suffices to prove that $X^2 + Y^2 - 1$ is irreducible in $K[Y][X]$ applying the Eisenstein criterion (see 17.21). Indeed, $Y + 1$ divides $Y^2 - 1$ but $(Y + 1)^2$ does not divide $Y^2 - 1$ because otherwise $Y^2 - 1 = (Y + 1)^2$ or $2Y + 2 = 0$. But this is impossible without $char(K) = 2$.

Ex. 5.16 (i) If $u \in K$ is a root for the polynomial f then $u^p = u + a$. But then $u + k, 1 \leq k \leq p - 1$ are also another $p - 1$ different roots for f and so f factors completely in $K[X]$. Indeed, $(u + 1)^p = u^p + \binom{p}{1} u^{p-1} + .. + \binom{p}{p-1} u + 1 =$

$u^p + 1 = (u + a) + 1 = (u + 1) + a$ shows that $u + 1$ is also a root $(char(K) = p$ is used) and one continues in this manner for $u + 2, .., u + (p - 1)$.

(ii) Set $x = by$ and use (i);

(iii) Set $x = z^{-1}$ and use (ii).

Ex. 5.17 First observe that $a \neq -1$: otherwise the existence of the map f would imply $char(K) = 2$. Indeed, for $a = -1$, $f(x) + f(-x) = 2x$ and replacing x by $-x$, $f(-x) + f(x) = -2x$ so that $2[f(x) + f(-x)] = 0$ or (multiplying by 2^{-1}) $f(x) + f(-x) = 2x = 0, \forall x \in K$.

Further, from $a^{2n+1} + 1 = (a + 1)(a^{2n} - a^{2n-1} + .. - a + 1)$ follows $1 - a + .. - a^{2n-1} + a^{2n} = 2(a + 1)^{-1}$ (because $a + 1 \neq 0$).

Finally, one checks that the map $f : K \to K, f(x) = 2(a + 1)^{-1} x$,

$\forall x \in K$ is the required one.

Remark. It can be shown that this is the unique map with the required properties.

Chapter 6

Divisibility in Integral Domains

Ex. 6.1 If $a \neq 0$ is an element of D then by hypothesis a and a^2 must be associated elements of D. Then $a^2 = au$ for a suitable unit $u \in D$. Hence $a(a - u) = 0$ and $a - u = 0$, D having no zero divisors. So, each nonzero element in D is a unit.

Ex. 6.2 We prove this by induction on the values of N. If $N(x) = 2$ and $x = a_1 a_2$ then $N(x) = N(a_1)N(a_2) = 2$ implies $N(a_1) = 1$, that is a_1 is a unit and a_2 is associated with x, or, $N(a_1) = 2$ and then a_1 is associated with x. Further, if $N(x) = n+1$ suppose that x is reducible. There are a_1, a_2 such that $x = a_1 a_2$ and $N(x) = N(a_1).N(a_2)$ and $N(a_1) \neq 1$. But then $N(a_1), N(a_2) < n+1$ so that by the induction hypothesis a_1, a_2 decompose into irreducible factors. Hence x decomposes into irreducible factors.

Ex. 6.3 It suffices to define a norm as follows:
$$N(x) = \begin{cases} 0 & \text{if } x = 0 \\ 1 & \text{for each unit } x \\ 2^{r_1 + .. + r_k} & \text{otherwise} \end{cases}$$
that is, if $x = \varepsilon p_1^{r_1} .. p_k^{r_k}$ with ε a unit and p_j irreducible elements. One verifies that N is well-defined and multiplicative.

Ex. 6.4 Let us define $N : \mathbb{Z}[\sqrt{n}] \to \mathbb{N}, N(a + b\sqrt{n}) = |a^2 - nb^2|$. If $\overline{a + b\sqrt{n}}$ denotes $a - b\sqrt{n}$ then $\overline{z.u} = \overline{z}.\overline{u}$ and $N(x) = |x.\overline{x}|$ simplify the verifications of the multiplicativity of N. The other two properties are immediate. For $\mathbb{Z}[i\sqrt{n}]$ we define the norm by $N(a + bi\sqrt{n}) = a^2 + nb^2$. The verifications are similar.

Ex. 6.5 Let $N : \mathbb{Z}[i]^* \to \mathbb{N}^*$, $N(a + bi) = a^2 + b^2$. If $z = a + bi \in \mathbb{Z}[i]$ and $w = c + di \in \mathbb{Z}[i]^*$ then let $z \cdot w^{-1} = s + ti \in \mathbb{Q}[i]$. If $q = m + ni \in \mathbb{Z}[i]$ is chosen such that $|s - m| \leq \frac{1}{2}, |t - n| \leq \frac{1}{2}$ then let $u = s - m + (t - n)i$. Hence $z = wq + wu$. Now, $z, wq \in \mathbb{Z}[i]$ and hence also $r = wu \in \mathbb{Z}[i]$. Moreover, $N(r) = N(wu) = N(w).N(u) = N(w).\left[(s - m)^2 + (t - n)^2\right] \leq \frac{1}{2}N(w) < N(w)$.

Ex. 6.6 One has $N(5) = 25$ but $a^2 + 2b^2 = 5$ has no solutions in $\mathbb{Z}$ so that 5 is irreducible. We have $2 + i\sqrt{2} = (1 - i\sqrt{2})(i\sqrt{2}) = (-1 + i\sqrt{2})(-i\sqrt{2})$ as non-trivial decomposition but $N(1 + i\sqrt{2}) = 3$ prime number, shows that this is an irreducible element so that one has only associate decompositions $1 + i\sqrt{2} = (-1)(-1 - i\sqrt{2})$.

Ex. 6.7 $\mathbb{Z}[i]$ being an Euclidean ring, one has to apply an Euclidean algorithm in order to get a gcd. First we obtain $5 = (1 + 3i)(-i) + 2 + i$, searching the quotient as a Gauss integer $m + ni$ such that $\left|\frac{1}{2} - m\right| \leq \frac{1}{2}, \left|-\frac{3}{2} - n\right| \leq \frac{1}{2}$ because $\frac{5}{1 + 3i} = \frac{1}{2} - \frac{3}{2}i$ so that $m = 0, n = -1$. Further, $1 + 3i = (2 + i)(1 + i)$ so that $\gcd(5, 1 + 3i) = 2 + i$.

Ex. 6.8 Let $N : \mathbb{Z}[i\sqrt{5}] \to \mathbb{N}$ be the function defined by $N(a + bi\sqrt{5}) = a^2 + 5b^2$. One verifies $N(xy) = N(x).N(y), \forall x, y \in \mathbb{Z}[i\sqrt{5}]$ so that if d would be a $\gcd(6, 2(1 + i\sqrt{5})$ then $N(d)$ divides $N(6) = 36$ and $N(2(1 + i\sqrt{5})) = 24$. But 2 divides $2(1 + i\sqrt{5})$ and $1 + i\sqrt{5}$ divides 6 (indeed $(1 + i\sqrt{5})(1 - i\sqrt{5}) = 6$) so that $N(2) = 4$ and $N(1 + i\sqrt{5}) = 6$ divide $N(d)$. Hence $N(d) = 12$. But there are no integers $a, b \in \mathbb{Z}$ such that $a^2 + 5b^2 = 12$.

Ex. 6.9 If $d = \gcd(3, 1 + i\sqrt{5})$, as above $N(d)$ divides $N(3) = 9$ and $N(1 + i\sqrt{5}) = 6$ so that $N(d) \in \{1, 3\}$. There are no $a, b \in \mathbb{Z}$ such that $a^2 + 5b^2 = 3$ so that $N(d) = 1$ and hence d is a unit (see 1.18) in $\mathbb{Z}[i\sqrt{5}]$. Hence $1 = \gcd(3, 1 + i\sqrt{5})$.

Ex. 6.10 $\mathbb{Z}[i]$ being an Euclidean ring a $\gcd(1+i, 4+i)$ will be the required generator so that we have to apply an Euclidean algorithm. We have $\frac{4+i}{1+i} = \frac{5}{2} - \frac{3}{2}i = 2 - i + \frac{1-i}{2}$ so (see an above solution) $4+i = (1+i)(2-i) + 1$ and hence $1 = \gcd(1+i, 4+i)$.

Ex. 6.11 We have $(1+i\sqrt{n})(1-i\sqrt{n}) = n + 1 = 2k$ (because $\sqrt{n} \notin \mathbb{Z}$, $n \geq 3$ surely implies that n is an odd number) so that $2|(1+i\sqrt{n})(1-i\sqrt{n})$. Simple verifications show that 2 does not divide neither $1 + i\sqrt{n}$ nor $1 - i\sqrt{n}$. So 2 is no prime element in $\mathbb{Z}[i\sqrt{n}]$. Further, in order to show that 2 is irreducible, let $2 = xy$ with $x, y \in \mathbb{Z}[i\sqrt{n}]$, namely $x = a + bi\sqrt{n}, y = c + di\sqrt{n}$, with $a, b, c, d \in \mathbb{Z}$. Using the conjugates $\bar{x}, \bar{y}$ we obtain $(a^2 + nb^2)(c^2 + nd^2) = 4$ so that $a^2 + nb^2 \in \{1, 2, 4\}$ respectively $a^2 + nb^2 \in \{1, 2\}$ (using symmetrically $c^2 + nd^2$). The first case implies $b = 0, a \in \{\pm 1\}$ and hence x is a unit. The second case is impossible.

Ex. 6.12 The ring $\mathbb{Z}[i]$ being Euclidean (see 6.5 above) the prime and the irreducible elements coincide so it is sufficient to show that $1 + 2i$ is irreducible. As we have seen in previous exercises it suffices to show that $N(1 + 2i)$ is prime (indeed $N(x) = N(y)$ implies x, y are associated). But $N(1 + 2i) = 5$.

Ex. 6.13 (a) According to 6.4 $\mathbb{Z}\left[i\sqrt{6}\right]$ is normed, by $N : \mathbb{Z}\left[i\sqrt{6}\right] \to \mathbb{N}$, $N\left(a + bi\sqrt{6}\right) = a^2 + 6b^2$. It is readily checked that $a^2 + 6b^2 \in \{2, 5\}$ have no solutions for $(a, b) \in \mathbb{Z} \times \mathbb{Z}$.

(b) For instance, $a^2 + 6b^2 = N\left(2 - i\sqrt{6}\right) = 10$ with $a, b \in \mathbb{Z}$ implies $a \in \{-2, 2\}$ and $b \in \{-1, 1\}$ so that $2 - i\sqrt{6}$ is irreducible. The same judgement is to be made for $N(2) = 4$ and $N(5) = 25$. Next, $6 = 2.3 = i\sqrt{6} \cdot \left(-i\sqrt{6}\right)$ are nonassociated decompositions and $2|6 \neq 2|i\sqrt{6}$ nor $2|\left(-i\sqrt{6}\right)$, so that neither 2 nor 3 is a prime element. For the third element one uses $\left(2 + i\sqrt{6}\right)\left(2 - i\sqrt{6}\right) = 2.5$

(c) We have $N\left(2 + i\sqrt{6}\right) = 10$ and $N(5) = 25$ so that $a|2 + i\sqrt{6}$ and $a|5$ imply $N(a)|\gcd(10, 25)$ that is $N(a)|5$. Hence, using (a), only $N(a) = 1$ is possible. Surely, $\left(2 + i\sqrt{6}, 5\right) \neq \mathbb{Z}\left[i\sqrt{6}\right]$ and so $1 \notin \left(2 + i\sqrt{6}, 5\right)$.

(d) Similarly $N(10) = 100, N\left(4 + 2i\sqrt{6}\right) = 40$ so that if $d = $ gcd $\left(10, 4 + 2i\sqrt{6}\right)$ we would have $d|10$ and $d|4 + 2i\sqrt{6}$ and so $N(d)|20$. But $2|10$ and $2|4 + 2i\sqrt{6}$ and so $N(2) = 4|N(d)$. Analogous, $2 + i\sqrt{6}|10$ and $2 + i\sqrt{6}|4 + 2i\sqrt{6}$ imply $10|N(d)$. Hence $N(d) = 20$ but $a^2 + 6b^2 = 20$ has no solutions in $\mathbb{Z}$.

(e) We have already seen that $10 = 2.5 = \left(2 + i\sqrt{6}\right).\left(2 - i\sqrt{6}\right)$ holds.

Ex. 6.14 More generally, we prove that in a Gaussian ring R which is not a field with only a finite number of units there are infinitely many irreducible elements. Indeed, suppose that $p_1, p_2, .., p_n$ would be the only nonassociated irreducible elements in R and consider the elements $q_k = (p_1 p_2 .. p_n)^k + 1$ for every $k \in \mathbb{N}$. If $k \neq l$ then $q_k \neq q_l$ otherwise the elements p_i would be units (if $k < l$ one has $(p_1 p_2 .. p_n)^{l-k} = 1$). The ring R having only a finite number of units there is a $r \in \mathbb{N}^*$ such that q_r is not a unit. If $q_r = u p_1^{r_1} p_2^{r_2} .. p_n^{r_n}$ is a decomposition in irreducible elements (with at least a $r_s > 0, 1 \leq s \leq n$) a simple computation leads to the conclusion $p_s|1$ or p_s is a unit, a contradiction.

Ex. 6.15 If an element $a \in D^*$ would not be a unit, the element a^2 would neither be a unit. By hypothesis, being irreducible, a and a^2 would be associated elements in D. But this implies a is a unit (see the solution of 6.1). This contradiction shows that D is a field.

Ex. 6.16 As above we use the function $N : \mathbb{Z}[i\sqrt{n}] \to \mathbb{N}, N(a + i\sqrt{n}) = a^2 + nb^2$ for $n \in \{3, 6\}$ in order to prove that $4 = 2.2 = (1 + i\sqrt{3})(1 - i\sqrt{3})$ respectively $6 = 2.3 = (i\sqrt{6})(-i\sqrt{6})$ are non-associated decompositions in irreducible elements. Indeed, in the first case we verify that neither 2 divides $1 \pm i\sqrt{3}$ nor $1 \pm i\sqrt{3}$ divides 2, and in the second case we recall that if $N(x) \neq N(y)$ then x and y are not associated.

Ex. 6.17 For $\mathbb{Z}[i\sqrt{2}]$ we prove a division algorithm using the corresponding function $N : \mathbb{Z}[i\sqrt{2}] \to \mathbb{N}, N(a + bi\sqrt{2}) = a^2 + 2b^2$ as follows: let $x = a + bi\sqrt{2}, y = c + di\sqrt{2} \neq 0$ and $\frac{x}{y} = u + vi\sqrt{2}$ with $u, v \in \mathbb{Q}$ and let $s, t \in \mathbb{Z}$ such that $|u - s| \leq \frac{1}{2}$ and $|v - t| \leq \frac{1}{2}$ (the nearest integers for u, v). We have $x = yq + r$ with $N(r) < N(y)$ for

$q = s + ti\sqrt{2}$ and $r = [(u-s)+(v-t)i\sqrt{2}].y \in \mathbb{Z}[i\sqrt{2}]$. Indeed, $N\left(\frac{r}{y}\right) = N\left(\frac{x}{y} - (s + ti\sqrt{2})\right) = (u-s)^2 + 2(v-t)^2 \leq \frac{1}{4} + 2.\frac{1}{4} = \frac{3}{4} < 1$. So the answer is affirmative.

Ex. 6.18 As above we consider $N : \mathbb{Z}\left[\sqrt{3}\right]^* \to \mathbb{N}^*, N\left(a + b\sqrt{3}\right) = |a^2 - 3b^2|$. If $a + b\sqrt{3}, c + d\sqrt{3}$ are nonzero elements it is readily checked that $N\left(a + b\sqrt{3}\right) \leq N\left[\left(a + b\sqrt{3}\right)\left(c + d\sqrt{3}\right)\right]$. Using the standard division algorithm in $\mathbb{Z}$ we obtain $ac - 3bd = q\left(c^2 - 3d^2\right) + r$ and $bc - ad = q'\left(c^2 - 3d^2\right) + r'$ such that $|r| \leq \frac{1}{2}|c^2 - 3d^2|, |r'| \leq \frac{1}{2}|c^2 - 3d^2|$. Now, if we consider $s + t\sqrt{3} = \left(a + b\sqrt{3}\right) - \left(q + q'\sqrt{3}\right)\left(c + d\sqrt{3}\right)$ then we have the required division algorithm in $\mathbb{Z}\left[\sqrt{3}\right]$ because $s + t\sqrt{3} = 0$ or $N\left(s + t\sqrt{3}\right) < N\left(c + d\sqrt{3}\right)$.

Chapter 7

Division Rings

Ex. 7.1 Let $K = \{0, 1, a, b\}$ be a field with four elements. Its characteristic being 2 (a prime number dividing 4) we obtain $1 + 1 = a + a = b + b = 0$. The group $(K, +)$ being of Klein type we also have $a + b = 1$. The inverse of a must be b. Indeed, $a^{-1} = 1$ would imply $a = 1$, $a^{-1} = a$ would imply $a^2 = 1$ and $(a-1)^2 = 0$ ($char(K) = 2$) and hence $a = 1$. So $a^{-1} = b$ or $ab = 1$ and even $a(1 + a) = 1$ or $a^2 = 1 + a$. An isomorphism to K_4 (from 1.9) is now obviously defined.

Ex. 7.2 One checks that this is a ring with identity. Let us denote by $a = (a, \overline{0}, \overline{0})$ and by $x = (\overline{0}, \overline{1}, \overline{0})$. Hence the 8 elements are $0, 1, x, 1 + x, x^2, 1 + x^2, x + x^2, 1 + x + x^2$. From $x^3 = 1 + x$ we infer $x(1 + x^2) = 1$ and so x and $1 + x^2$ are units. Moreover, $x^2(1 + x + x^2) = 1$ and $(1 + x)(x + x^2) = 1$ show that all the nonzero elements in K_8 are units.

Ex. 7.3 As in the previous exercise we denote by $a = (a, \overline{0})$ and by $y = (\overline{0}, \overline{1})$. The 9 elements are $0, 1, 2, y, 2y, 1 + y, 1 + 2y, 2 + y, 2 + 2y$ and the equalities $y^2 = 2, y \cdot 2y = 1, (1 + y)(2 + y) = 1, (1 + 2y)(2 + 2y) = 1$ show that all the nonzero elements in K_9 are units.

Ex. 7.4 The map $f : \mathbb{C} \rightarrow \left\{ \begin{pmatrix} a & b \\ -b & a \end{pmatrix} \middle| a, b \in \mathbb{R} \right\}$ defined by $f(a + bi) = \begin{pmatrix} a & b \\ -b & a \end{pmatrix}$ is the required ring isomorphism.

Ex. 7.5 The verifications are easier if one uses the decomposition

$$M(x,y) = \begin{pmatrix} x+2y & 3y \\ 2y & x-2y \end{pmatrix} = xI_2 + yA \text{ where } I_2 \text{ denotes the}$$

identity matrix and $A = \begin{pmatrix} 2 & 3 \\ 2 & -2 \end{pmatrix}$. We note that $A^2 = 10I_2$ and then

$M(x,y)+M(z,t) = M(x+z,y+t)$, $M(x,y).M(z,t) = M(xz+10yt, xt+yz)$. Moreover, if $M(x,y) \neq M(0,0)$ we have $x^2 - 10y^2 \neq 0$ ($\sqrt{10}$ being not rational) and $M(x,y)^{-1} = M\left(\frac{x}{x^2-10y^2}, -\frac{y}{x^2-10y^2}\right)$. The map $f: \mathbb{Q}[\sqrt{10}] \to \{M(x,y)|x,y \in \mathbb{Q}\}$ defined by $f(x+y\sqrt{10}) = M(x,y)$ is the required ring isomorphism.

Ex. 7.6 As above one simplifies the verifications decomposing $\begin{pmatrix} a & b \\ db & a \end{pmatrix} = aI_2+bA$, where $A = \begin{pmatrix} 0 & 1 \\ d & 0 \end{pmatrix}$ and using $A^2 = dI_2$. The required isomorphism is defined by $a + b\sqrt{d} \longmapsto \begin{pmatrix} a & b \\ db & a \end{pmatrix}$.

Ex. 7.7 Let $f : (K,+) \to (K^*,\cdot)$ be a group isomorphism. Then $f(0) = 1$ and there is an element $a \in K^*$ such that $f(a) = -1$. Now $f(-a) = f(0-a) = f(0) \cdot (f(a))^{-1} = 1 \cdot (-1)^{-1} = -1$ and so, f being injective, $a = -a$ or $2a = 0$. But then $(2 \cdot 1) \cdot a = 0$ and hence $2 \cdot 1 = 0$ which implies $2x = 0, \forall x \in K$. Finally $1 = f(0) = f(2 \cdot 1) = (f(1))^2$ or $(f(1))^2 - 1 = (f(1) - 1)^2 = 0$. Hence $f(1) = 1$ and using the injectivity of f, $0 = 1$, a contradiction. (see also the following exercise).

Ex. 7.8 If $char(K) < \infty$ then $char(K) = p$, a prime number. The field being a commutative ring, the Newton binomial formula gives (see 8.11) $(a \pm b)^p = a^p \pm b^p, \forall a, b \in K$. Hence $[f(x)]^p = f(px) = f(0) = 1, \forall x \in K$ and then $[f(x)]^p - 1^p = [f(x) - 1]^p = 0$. But so $f(x) = 1, \forall x \in K$ and then f is the trivial group homomorphism. This result cannot be applied to $\mathbb{Q}$ because $char(\mathbb{Q}) = \infty$. But this result remains true: there is only the trivial group homomorphism between $(\mathbb{Q},+)$ and $(\mathbb{Q}^*,\cdot)$. Indeed, for every $x \in \mathbb{Q}$ and a such group homomorphism f from $f(x) = f\left(\frac{x}{2} + \frac{x}{2}\right) = \left[f\left(\frac{x}{2}\right)\right]^2 > 0$. Next, one checks $f(x) = [f(1)]^x, \forall x \in \mathbb{Q}$. Finally, we prove $f(1) = 1$. If $f(1) = \frac{a}{b}$ with $a, b \in \mathbb{N}^*, a \neq b, (a;b) = 1$, choosing $p \in \mathbb{N}, p > \max(a,b)$, at least one from a and b is not the

p-th power of a natural number and hence $\frac{a}{b} \notin (\mathbb{Q})^p$. Hence, $f\left(\frac{1}{p}\right) = [f(1)]^{\frac{1}{p}} = \left(\frac{a}{b}\right)^{\frac{1}{p}} \in \mathbb{Q}$, a contradiction.

Ex. 7.9 Multiplying to the right the first relation by a^n one obtains $a^n b^n a^n - b^{n+1} a^{2n+1} = a^n$. Further, multiplying by b^{n+1} to the left and then to the right $a^{2n+1} = -b^{2n+1}$ we get $b^{n+1} a^{2n+1} = -b^{3n+2}$,

$a^{2n+1} b^{n+1} = -b^{3n+2}$ and hence $b^{n+1} a^{2n+1} = a^{2n+1} b^{n+1}$. Using a previous relation we deduce $a^n b^n a^n - a^{2n+1} b^{n+1} = a^n$ and hence the required relation follows by left multiplication with a^{-n}.

Ex. 7.10 Obviously, only the "if" part needs verification. We show that if $xy = 1$ then also $yx = 1$ holds. If $xy = 1$ then $y \neq 0$ so that there is an element $z \in R$ such that $yz = 1$. But $z = 1.z = (xy).z = x(yz) = x.1 = x$ so that $yx = 1$ as required.

Remark. One surely obtains a similar true statement replacing the existence of the right inverses by the corresponding left inverses.

Ex. 7.11 In a division ring, if $x \neq 1$, the equation $x + y = xy$ has the unique solution $y = (x - 1)^{-1}x$. Conversely, for an arbitrary element $a \in R, a \neq 0$ we have $1 - a \neq 1$ so that there is a $b \in R$ such that $1 - a + b = (1 - a)b$ and hence $1 = a(1 + b)$, i.e. a has a right inverse. By the above exercise, R is a division ring.

Ex. 7.12 In what follows we give three different (but isomorphic) presentations of the so-called *quaternions* division ring.

1. $\mathbf{H} = \mathbb{R}^4$ together with the following operations $(a, b, c, d) + (a', b', c', d') = (a + a', b + b'c + c', d + d')$ and $(a, b, c, d).(a', b', c', d') = (aa' - bb' - cc' - dd', ab' + ba' + cd' - dc', ac' - bd' + ca' + db', ad' + bc' - cb' + da')$. Denoting by $\mathbf{i} = (0, 1, 0, 0), \mathbf{j} = (0, 0, 1, 0), \mathbf{k} = (0, 0, 0, 1)$ and by $a = (a, 0, 0, 0), \forall a \in \mathbb{R}$ one gets the representation $(a, b, c, d) = a + b\mathbf{i} + c\mathbf{j} + d\mathbf{k}$ together with $a\mathbf{i} = \mathbf{i}a, a\mathbf{j} = \mathbf{j}a, a\mathbf{k} = \mathbf{k}a, \mathbf{i}^2 = \mathbf{j}^2 = \mathbf{k}^2 = -1, \mathbf{i}\mathbf{j} = \mathbf{k} = -\mathbf{j}\mathbf{i}, \mathbf{j}\mathbf{k} = \mathbf{i} = -\mathbf{k}\mathbf{j}, \mathbf{k}\mathbf{i} = \mathbf{j} = -\mathbf{i}\mathbf{k}$ which simplifies all the needed verifications. If $n = a^2 + b^2 + c^2 + d^2 \neq 0$ then $(a, b, c, d)^{-1} = \left(\frac{a}{n}, -\frac{b}{n}, -\frac{c}{n}, -\frac{d}{n}\right)$.

2. $H = \left\{ \begin{pmatrix} a & b & c & d \\ -b & a & -d & c \\ -c & d & a & -b \\ -d & -c & b & a \end{pmatrix} \,|\, a, b, c, d \in \mathbb{R} \right\} \subseteq M_4(\mathbb{R})$ to-

gether with the usual matrices addition and multiplication. The astute reader will simplify the verifications decomposing

$$\begin{pmatrix} a & b & c & d \\ -b & a & -d & c \\ -c & d & a & -b \\ -d & -c & b & a \end{pmatrix} = aI_4 + bJ + cK + dL \text{ where}$$

$$J = \begin{pmatrix} 0 & 1 & 0 & 0 \\ -1 & 0 & 0 & 0 \\ 0 & 0 & 0 & -1 \\ 0 & 0 & 1 & 0 \end{pmatrix}, K = \begin{pmatrix} 0 & 0 & 1 & 0 \\ 0 & 0 & 0 & 1 \\ -1 & 0 & 0 & 0 \\ 0 & -1 & 0 & 0 \end{pmatrix} \text{ and}$$

$$L = \begin{pmatrix} 0 & 0 & 0 & 1 \\ 0 & 0 & -1 & 0 \\ 0 & 1 & 0 & 0 \\ -1 & 0 & 0 & 0 \end{pmatrix} \text{ with similar relations to the above ones}$$

given for the quaternions $\mathbf{i}, \mathbf{j}, \mathbf{k}$. A natural map gives $\mathbf{H} \cong H$.

3. $\mathcal{H} = \left\{ \begin{pmatrix} z & u \\ -\bar{u} & \bar{z} \end{pmatrix} \,|\, z, u \in \mathbb{C} \right\} \subseteq M_2(\mathbb{C})$ where $\bar{z}$ denotes the

conjugate of z, together with the usual matrices addition and multipli-

cation. The map $\begin{pmatrix} z & u \\ -\bar{u} & \bar{z} \end{pmatrix} \longmapsto a + b\mathbf{i} + c\mathbf{j} + d\mathbf{k}$, where $z = a + b\mathbf{i}, u =$

$c + d\mathbf{i}$ provides $\mathcal{H} \cong \mathbf{H}$.

Ex. 7.13 For a prime number p consider $K = \mathbb{Z}_p[X]/(X^2 + aX + b)$. If the polynomial $X^2 + aX + b$ is irreducible over $\mathbb{Z}_p$ then the ideal $(X^2 + aX + b)$ is maximal and K is a field with p^2 elements (see also 7.15 bellow). In order to prove that such a polynomial ($\mathbb{Z}_p$ being a field it suffices to consider monic polynomials, i.e. polynomials with the leading coefficient $\bar{1}$) does exist it suffices to make the following simple computation: there are obviously only p monic polynomials of degree 1 (that is $X, X + \bar{1}, ..X + \overline{p-1}$) and so there are $\binom{p}{2} + p = \frac{p(p+1)}{2}$ reducible monic polynomials of degree 2. Hence there are $p^2 - \frac{p(p+1)}{2} = \frac{p(p-1)}{2}$ irreducible polynomials of degree 2.

Ex. 7.14 We compute the center of $\mathbf{H}$, the division ring of all the quaternions. Let $q = a + bi + cj + dk \in Z(\mathbf{H})$. From $q \cdot i = i \cdot q$ we infer $ai - b - ck + dj = ai - b + ck - dj$ and so $c = d = 0$. From $q \cdot j = j \cdot q$ we deduce $b = 0$. Hence $Z(\mathbf{H}) = \mathbb{R}$.

Ex. 7.15 K being finite, it has a finite and hence prime characteristic p. Then K contains the field $\mathbb{Z}_p$ and moreover, is a linear space over $\mathbb{Z}_p$. Indeed, if one uses 5.6, $char(K) = p$ implies that for f, the unique unital ring homomorphism, $f(\mathbb{Z}) \cong \mathbb{Z}/\ker(f) = \mathbb{Z}/(p) \cong \mathbb{Z}_p$. If $\dim_{\mathbb{Z}_p}(K) = n$ then K has p^n elements.

Ex. 7.16 Nonzero endomorphisms of fields are injective and hence (the restriction to K^* is homomorphism of groups) unital. So $0, 1 \in H$ and so $|H| \geq 2$. Moreover, it is readily checked that $a, b \in H, c \in H \backslash \{0\} \Rightarrow a - b \in H$ and $a.c^{-1} \in H$ (indeed, $f(c^{-1}) = f(c)^{-1}$).

Ex. 7.17 .Let $N = \{1, 2, .., n\}$ and $M = \{1, 2, ..m\}$ for $n, m \in \mathbb{N}^*$. Then if we denote by $E_{ij} = U_{ij}(1)$ (the matrix whose $i - j$ entry is 1, with all the other entries 0) then $\mathcal{M}_n(K) = \{f | f : N \times N \to K\}$ and $\mathcal{M}_m(K') = \{g | g : M \times M \to K'\}$ (all the maps) can be considered as linear K-space resp. K'-space with bases $\{E_{ij} | i, j \in N\}$ resp. $\{E'_{ij} | i, j \in M\}$. Now $K \cong E_{11} \mathcal{M}_n(K) E_{11}$ (notice that E_{11} is obviously an idempotent) and $K' \cong E'_{11} \mathcal{M}_m(K') E'_{11}$ together with an isomorphism $F : \mathcal{M}_n(K) \to \mathcal{M}_m(K')$ gives by restriction an isomorphism $K \cong K'$. To get this, notice that, for instance $\{E_{ij} | i, j \in N\}$ (called *matric units* in [25]) are determined as elements in a ring by identities preserved by a ring isomorphism : $\sum_{i=1}^{n} E_{ii} = I_n, E_{ij} E_{kl} = \delta_{jk} E_{il}$.

Finally, if $K \cong K'$ then $\dim_K(\mathcal{M}_n(K)) = n^2 = m^2 = \mathcal{M}_m(K')$ implies $n = m$.

Chapter 8

Automorphims

Ex. 8.1 If $f : \mathbb{Q} \to \mathbb{Q}$ is an automorphism one easily checks that $f(0) = 0$ and $f(1) = 1$. Then for each $n \in \mathbb{N}^*$ we deduce $f(n) = f(1+..+1) = f(1) + .. + f(1) = nf(1) = n$. Moreover, $f(-n) = -f(n) = -n$ so that $f(m) = m, \forall m \in \mathbb{Z}$. Further, for each $n \in \mathbb{Z}^*, 1 = f(1) = f\left(n \cdot \frac{1}{n}\right) = f(n) \cdot f\left(\frac{1}{n}\right) = nf\left(\frac{1}{n}\right)$ and hence $f\left(\frac{1}{n}\right) = \frac{1}{n}$. Finally, for each $\frac{m}{n} \in \mathbb{Q}$ we have $f\left(\frac{m}{n}\right) = f\left(m \cdot \frac{1}{n}\right) = f(m) \cdot f\left(\frac{1}{n}\right) = m \cdot \frac{1}{n} = \frac{m}{n}$.

Ex. 8.2 From the previous exercise we know that $\mathbb{Q}$ has only the identical automorphism. Now, if $f : \mathbb{R} \to \mathbb{R}$ is an automorphism, as above $f(a) = a, \forall a \in \mathbb{Q}$. If $x \in \mathbb{R}$ and $x > 0$ then $f(x) = f((\sqrt{x})^2) = (f(\sqrt{x}))^2 > 0$ ($f(0) = 0$ and the injectivity of f imply $f(\sqrt{x}) \neq 0$). Taking $x_1 < x_2$ and $x = x_2 - x_1$ we infer $f(x_1) < f(x_2)$ so that f is an increasing function. We prove that f is the identical automorphism. If we would have $f(x) \neq x$ for a suitable $x \in \mathbb{R}$ suppose for example $f(x) < x$. There is an $a \in \mathbb{Q}$ such that $f(x) < a < x$ so that $a = f(a) < f(x)$ is a contradiction. Finally, if $f : \mathbb{C} \to \mathbb{C}$ is such that $f(x) = x, \forall x \in \mathbb{R}$ then $-1 = f(-1) = f(i^2) = (f(i))^2$ so that $f(i) \in \{-i, i\}$. Hence f is the identical automorphism or the conjugacy of complex numbers, i.e. $f(a + bi) = a - bi$ for every $a + bi \in \mathbb{C}$.

Ex. 8.3 Let $f : \mathbb{Q}(\sqrt{d}) \to \mathbb{Q}(\sqrt{d})$ be an automorphism. As in the previous exercises from $f(0) = 0$ and $f(1) = 1$ we deduce $f\left(\frac{m}{n}\right) = \frac{m}{n}, \forall \frac{m}{n} \in \mathbb{Q}$. Now, for each $a, b \in \mathbb{Q}$, $f(a+b\sqrt{d}) = f(a)+f(b)f(\sqrt{d}) = a+$

$bf(\sqrt{d})$ holds so that f is determined by $f(\sqrt{d})$. If $f(\sqrt{d}) = x + y\sqrt{d}$ the following computation is needed: $d + f(d) = f(\sqrt{d} \cdot \sqrt{d}) = (f(\sqrt{d}))^2 = (x + y\sqrt{d})^2 = x^2 + dy^2 + 2xy\sqrt{d}$. It implies $x^2 + dy^2 = d, xy = 0$. If $y = 0$ one would have $d = x^2$ and hence d would not be square-free. So $x = 0$ and $y \in \{\pm 1\}$. Hence $\mathbb{Q}[\sqrt{d}]$ has only two automorphisms (the identical and "the conjugation" ones).

Ex. 8.4 For each automorphism $f : K_4 \to K_4$ we have $f(0) = 0$ and $f(1) = 1$. The only bijective functions verifying these conditions are the identical automorphism and the function f such that $f(a) = b, f(b) = a$.

Ex. 8.5 We use the representation pointed out in 7.2:

$K_8 = \{0, 1, x, 1 + x, x^2, 1 + x^2, x + x^2, 1 + x + x^2\}$. Because $f(0) = 0, f(1) = 1$, $f(x)$ (which obviously determines an automorphism f of K_8) may be chosen from the other 6 elements. But one verifies that $x^3 = 1 + x$ holds and hence also $f(x)^3 = 1 + f(x)$. Straightforward computations show that this holds only for the elements $x, x^2, x + x^2$. So K_8 has only 3 automorphisms: if we denote by $\alpha : K_8 \to K_8$ the automorphism determined by $\alpha(x) = x^2$ then $\alpha^3 = 1_{K_8}$ and the 3 automorphisms are the identical one and α, α^2.

Ex. 8.6 We use the representation pointed out in 7.3:

$K_9 = \{0, 1, 2, y, 2y, 1 + y, 1 + 2y, 2 + y, 2 + 2y\}$. Because $f(0) = 0, f(1) = 1, f(2) = 2, f(y)$ (which determines an automorphism f of K_9) may be chosen from the other 7 elements. Now, one verifies that $y^2 = y + 1$ so that $(f(y))^2 = 1 + f(y)$. We easily check that this is possible only if $f(y) \in \{y, 1 + 2y\}$ so we have only a non-identical automorphism $\beta : K_9 \to K_9$, determined by $\beta(y) = 1 + 2y$.

Ex. 8.7 $\mathbb{Z}(\sqrt{3})$ is a subring of $\mathbb{Q}(\sqrt{3})$ group-generated by 1 and $\sqrt{3}$. One verifies that the only non-identical automorphism is $m + n\sqrt{3} \longmapsto m - n\sqrt{3}$.

Ex. 8.8 We use 4.8: the ring homomorphisms of $\mathbb{Z}_p$ are the translations with the idempotents. $\mathbb{Z}_p$ being a field, the only idempotent is $\overline{1}$ so that $\mathbb{Z}_p$ has only the identical automorphism.

Ex. 8.9 (a) If a is a unit, the map $\psi : R[X] \to R[X]$, $\psi(p(X)) = p(a^{-1}X - a^{-1}b)$ is the inverse of φ, so that straightforward verifications show that these are automorphisms of $R[X]$.

(b) Let $f : R[X] \to R[X]$ be an automorphism with inverse g such that $f(a) = a, \forall a \in R$. We have $(f \circ g)(X) = X$ so that $f(X)$ and $g(X)$ are polynomials of degree one. If $f(X) = a_1 X + a_0$ and $g(X) = b_1 X + b_0$ the previous equality gives $a_1 b_1 = 1$ and $a_1 b_0 + a_0 = 0$ so that a_1 is a unit. Hence f is one of the automorphisms pointed out in (a). Indeed, $f(p(X)) = f(c_0 + c_1 X + .. + c_n X^n) = f(c_0) + f(c_1)f(X) + .. + f(c_n)f(X)^n = p(f(X)) = p(a_1 X + a_0)$.

(c) From the above exercise, each automorphism φ of $\mathbb{Z}_p$ is defined by $\varphi(p(X)) = p(aX + b)$ where $a, b \in \mathbb{Z}_p$ and a is a unit. Hence $\mathbb{Z}_p[X]$ has $p(p - 1)$ automorphisms, $\mathbb{Z}_p$ having $p - 1$ units. Similarly, the cardinal number of $Aut(\mathbb{Q}[X])$ is $\aleph_0$.

Ex. 8.10 Indeed, first observe that each automorphism of $\mathbb{Q}(\sqrt[3]{a})$ maps $\sqrt[3]{a}$ into a root of the equation $x^3 = a$ (if $f(\sqrt[3]{a}) = b$ one has $b^3 = f(\sqrt[3]{a})^3 = f((\sqrt[3]{a})^3) = f(a) = a$ where the last equality is obtained as in the first exercise above). But the only real root of this equation is $\sqrt[3]{a}$ so that the only automorphism of $\mathbb{Q}(\sqrt[3]{a})$ is the identical one.

Ex. 8.11 K being commutative we first show that f is a ring homomorphism. Indeed, $f(ab) = (ab)^p = a^p b^p = f(a)f(b)$ and $f(a + b) = (a + b)^p = a^p + \binom{p}{1} a^{p-1}b + .. + \binom{p}{p-1} ab^{p-1} + b^p$. One has to know that a prime number p divides all the binomial coefficients $\binom{p}{1}, \binom{p}{2}, ..., \binom{p}{p-1}$. K having the characteristic p the above computation continues as follows: $f(a + b) = a^p + b^p = f(a) + f(b)$. Each non-trivial ring homomorphism between fields being injective, f is injective. K being finite, f is also surjective and hence, an automorphism.

Ex. 8.12 Let K be a division ring which has only a finite number of automorphisms. Among these, for each $0 \neq a \in K$ surely the maps $f_a : K \to K, f_a(x) = a^{-1}xa, \forall x \in K$ are automorphisms (called *inner*). Moreover, $(\{f_a | a \in K^*\}, \circ)$ has to be a finite group, a subgroup of

$Aut(K)$. Hence each element in this subgroup has finite order, i.e. for each $a \in K^*$ there is an $n(a) \in \mathbb{N}^*$ such that $(f_a)^{n(a)} = 1_K$ But then $(f_a \circ f_a \circ .. \circ f_a)(x) = x$ or $a^{-n(a)}xa^{n(a)} = x$ or $xa^{n(a)} = a^{n(a)}x, \forall x \in K$. Finally, for each $a \in K$ (clearly 0 also has this property) there is a number $n(a) \in \mathbb{N}^*$ such that $a^{n(a)} \in Z(K)$. But hence K is commutative according to 17.6 (obviously a division ring has no nilpotent ideals, excepting (0)).

Ex. 8.13 For each ring automorphism $f : R \to R$ one has $f(0) = 0, f(1) = 1$. But there are exactly $(n(R) - 2)!$ permutations of R with 0 and 1 fixed. Hence for the required rings each permutation of R which fixes 0 and 1 must be an automorphism.

First we show that $n(R) \leq 4$ and more precisely that $n(R) \geq 4 \Rightarrow n(R) = 4$. Consider the transposition $\sigma = (ab) \in S_{n(R)}$ where $a, b \in R \backslash \{0, 1\}$. As a permutation which fixes 0 and 1, σ must be an automorphism of R and consequently, also a group automorphism of $(R, +)$. Obviously, $\sigma \circ \sigma = 1_R$ and the verification reduces to $n(R) \geq 4 \Rightarrow n(R) = 4$ as (abelian) groups. Indeed, consider $r \in R \backslash \{0, a, b\}$ and analyse the sum $a + r$. If $a + r = 0$ applying σ, $b + r = 0$ follows and then $a = b$, a contradiction. The possibilities $a + r = a$ and $a + r = r$ clearly provide contradictions ($r = 0$ and $a = 0$). If $a + r = r' \in R \backslash \{0, a, b, r\}$ we apply σ and obtain $b + r = r'$ and $a = b$, a contradiction. Hence, only $a + r = b$ is possible and $x = b - a$ is uniquely determined. So $(R, +)$ has at most four elements. If one considers the permutation $\tau = \begin{pmatrix} \bar{0} & \bar{1} & \bar{2} & \bar{3} \\ \bar{0} & \bar{2} & \bar{1} & \bar{3} \end{pmatrix}$, by straightforward computation $\tau(\bar{2} + \bar{2}) = \tau(\bar{0}) = \bar{0} \neq \bar{2} = \bar{1} + \bar{1} = \tau(\bar{2}) + \tau(\bar{2})$ so that τ is not a group homomorphism. So $(R, +) \cong (\mathbb{Z}_3, +)$ or $(R, +) \cong (\mathbb{Z}_2 \times \mathbb{Z}_2, +)$ because $n(R) \geq 3$.

Using 1.9 (and 1.10) the only rings which could have the required property are $(\mathbb{Z}_3, +, \cdot)$ and the rings I_4, K_4 and $(\mathbb{Z}_2 \times \mathbb{Z}_2, +, \cdot)$ (the direct product) from 1.9. For these rings the subjacent abelian group is the Klein group $\mathcal{K} = \{0, 1, b, c\}$. The only permutations which fix 0 and 1 are $1_{\mathcal{K}}$ (which obviously is a ring homomorphism for all the structures) and the transposition $\delta = (bc)$. This transposition is a ring endomorphism for K_4 and the direct product but is not a ring endomorphism for I_4 : $\delta(bc) = \delta(b) = c \neq b = cb = \delta(b)\delta(c)$. Hence the required rings are $(\mathbb{Z}_3, +, \cdot)$, K_4 and $(\mathbb{Z}_2 \times \mathbb{Z}_2, +, \cdot)$.

Chapter 9

The Tensor Product

Ex. 9.1 Indeed, the map $f : \mathbb{Z} \otimes_{\mathbb{Z}} \mathbb{Z}_n \to \mathbb{Z} \cdot \mathbb{Z}_n$, $f(\sum_{i=1}^{n} x_i \otimes \bar{y}_i) = \sum_{i=1}^{n} x_i \bar{y}_i$
is readily seen to be a $\mathbb{Z}$-isomorphism.

More generally, if $_RM$ is a left R-module the map $\mu_R : R \otimes_R M \to RM$ defined as above (the corresponding map $R \times M \to M, (r, m) \mapsto rm$ being bilinear) is a R-epimorphism which is a R-isomorphism if R has identity ($RM = M$ and $\mu_R^{-1}(x) = 1 \otimes x$ is the inverse).

Ex. 9.2 (a) If $x \in \mathbb{Z}_n$ and $a \in \mathbb{Q}$ then $x \otimes a = x \otimes \left(\frac{1}{n}a\right) n = xn \otimes \frac{1}{n}a = 0 \otimes \frac{1}{n}a = 0$.

(b) Similarly $x \otimes (a + \mathbb{Z}) = x \otimes (\frac{1}{n}a + \mathbb{Z})n = 0 \otimes \left(\frac{1}{n}a + \mathbb{Z}\right) = 0$.

(c) $\left(\frac{m}{n} + \mathbb{Z}\right) \otimes a = \left(\frac{m}{n} + \mathbb{Z}\right) \otimes \left(\frac{1}{n}a\right) n = n \left(\frac{m}{n} + \mathbb{Z}\right) \otimes \frac{1}{n}a = \mathbb{Z} \otimes \frac{1}{n}a = 0 \otimes \frac{1}{n}a = 0$.

Remark. If R is an integral domain, $_RM$ is a torsion R-module and $_RN$ is a divisible R-module, more generally, one has $M \otimes_R N = 0$.

Ex. 9.3 The image of this homomorphism is zero: indeed,
$$u \otimes_{\mathbb{Z}} 1_{\mathbb{Z}}(nx, y) = nx \otimes y = x \otimes ny = x \otimes 0 = 0.$$
But $n\mathbb{Z} \otimes_{\mathbb{Z}} \mathbb{Z}_n \cong \mathbb{Z} \otimes_{\mathbb{Z}} \mathbb{Z}_n \cong \mathbb{Z} \neq 0$ (see also 9.1). Hence the homomorphism is not injective.

Ex. 9.4 We first consider the map $f : (R/I) \times M \to M/(I \cdot M)$,
$f(\bar{r}, x) = \overline{rx}$. Simple verifications show that f is well-defined (i.e. $s \in \bar{r} = r + I \Rightarrow sx \in \overline{rx} = rx + I \cdot M$) and R-bilinear. The definition of

the tensorial product gives a morphism of R-modules $\overline{f} : (R/I) \otimes_R M \to M/(I \cdot M)$, $\overline{f}(\overline{r} \otimes x) = \overline{rx}$.

Conversely, the map $h : M/(I \cdot M) \to (R/I) \otimes_R M$, $h(\overline{x}) = \overline{1} \otimes x$ is well-defined (i.e. $y \in \overline{x} = x + I \cdot M \Rightarrow \overline{1} \otimes x = \overline{1} \otimes y$) and homomorphism of R-modules. Finally, $\overline{f}$ and h are mutually inverse and so they provide the required isomorphism.

The application: *let G be an abelian group. There is a canonical isomorphism $\mathbb{Z}_m \otimes_\mathbb{Z} G \cong G/mG$ for each $m \in \mathbb{Z}$.*

Ex. 9.5 Using the previous exercise $R/I \otimes_R R/J \cong (R/I)/J \cdot (R/I) = (R/I)/(I + J/I) \cong R/(I + J)$ the only thing the reader has to verify is the equality $J \cdot (R/I) = (I + J)/I$.

Ex. 9.6 (a) Using the previous exercise we have $\mathbb{Z}_n \otimes \mathbb{Z}_m = \mathbb{Z}/n\mathbb{Z} \otimes \mathbb{Z}/m\mathbb{Z} \cong \mathbb{Z}/(n\mathbb{Z} + m\mathbb{Z}) = \mathbb{Z}/d\mathbb{Z} = \mathbb{Z}_d$

(b) If $\mathbb{Z}_n \otimes \mathbb{Z}_m = 0$ using again the previous exercise one gets $n\mathbb{Z} + m\mathbb{Z} = \mathbb{Z}$ and then $d = 1$. The converse follows from (a).

(c) Again using the previous isomorphism one has $\mathbb{Z}_n \otimes \mathbb{Z}_n \cong \mathbb{Z}/(n\mathbb{Z} + n\mathbb{Z}) = \mathbb{Z}/n\mathbb{Z} = \mathbb{Z}_n$

For each abelian group G and each $n \in \mathbb{N}^*$ one proves $\mathbb{Z}_n \otimes_\mathbb{Z} G \cong G/nG$.

Ex. 9.7 The map $m : \mathbb{Q} \times \mathbb{Q} \to \mathbb{Q}$, $m(a, b) = a.b, \forall a, b \in \mathbb{Q}$ is clearly $\mathbb{Z}$-bilinear so that there is an abelian groups homomorphism $f : \mathbb{Q} \otimes_\mathbb{Z} \mathbb{Q} \to \mathbb{Q}$, $f(a \otimes b) = ab, a, b \in \mathbb{Q}$. But the function $g : \mathbb{Q} \to \mathbb{Q} \otimes_\mathbb{Z} \mathbb{Q}$, $g(x) = 1 \otimes x, \forall x \in \mathbb{Q}$ is easily checked to be the inverse of f.

Ex. 9.8 Recall that for M_R and $_R N$ modules, a bilinear map $\varphi : M \times N \to G$ (where G is an abelian group) is called *universal* if for each bilinear map $\psi : M \times N \to H$ to an abelian group H, ψ factors through φ. The bilinear map $\otimes : M \times N \to M \otimes_R N$ is universal.

The map $\mu : K \times U \to U$, $\mu(k, u) = ku$ is easily seen to be bilinear so it induces a linear application $\overline{\mu} : K \otimes_K U \to U$, $\overline{\mu}(k, u) = k \otimes u$. Since $\overline{\mu}(1 \otimes u) = u, \forall u \in U$ this map is surjective. It is also injective: indeed, one establishes at once that each $x \in K \otimes_K U$ can be represented as $1 \otimes u$ for a suitable $u \in U$ (if $x = \sum_i (k_i \otimes u_i) = \sum_i (1k_i \otimes u_i) = \sum_i (1 \otimes k_i u_i) =$

$1 \otimes \left(\sum_i k_i u_i \right) = 1 \otimes u$). If therefore $f(x) = f(1 \otimes u) = 0$ then $u = 0$ and hence $x = 0$.

The above map $\otimes$ being bilinear, the map $\overline{\otimes} : U \times V \to V \otimes_K U, (u, v) \mapsto v \otimes u$, is also bilinear. By $\otimes$'s universality there is a unique linear map $t : U \otimes_K V \to V \otimes_K U$ such that $t(u \otimes v) = v \otimes u$. In order to prove that t is the required isomorphism it suffices to prove that $\overline{\otimes}$ is also universal (providing an inverse for t). Let $f : U \times V \to V \times U, f(u, v) = (v, u)$ be the obvious isomorphism. For each bilinear map $s : V \times U \to W$ there is a unique linear map h such that $h \circ \otimes \circ f^{-1} = s$ because of the bilinearity of $s \circ f$ and the universality of $\otimes$ (the above equality being equivalent to $h \circ \otimes = s \circ f$). Hence $\otimes \circ f^{-1} = \overline{\otimes}$ is universal.

For each $u \in U$ the map $\varphi_u : V \times W \to (U \otimes_K V) \otimes_K W, \varphi_x(v, w) = (u \otimes v) \otimes w$ is clearly bilinear and hence there is a unique homomorphism $\mu_u : V \otimes_K W \to (U \otimes_K V) \otimes_K W, \mu_u(v \otimes w) = (u \otimes v) \otimes w$. Further, we define $\varphi : U \times (V \otimes_K W) \to (U \otimes_K V) \otimes_K W$ by $\varphi(u, \sum_i (v_i \otimes w_i)) = \mu_u(\sum_i (v_i \otimes w_i)) = \sum_i (u \otimes v_i) \otimes w_i$ and this is also a bilinear map inducing the homomorphism $\mu : U \otimes_K (V \otimes_K W) \to (U \otimes_K V) \otimes_K W, \mu(u \otimes (\sum_i v_i \otimes w_i)) = \sum_i (u \otimes v_i) \otimes w_i$. In an analogous manner one gets an inverse for μ, i.e. $\mu' : (U \otimes_K V) \otimes_K W \to U \otimes_K (V \otimes_K W)$.

Ex. 9.9 If $\{u_i\}_{i \in I}$ and $\{v_j\}_{j \in J}$ are bases in U respectively V then $\{u_i \otimes v_j\}_{(i,j) \in I \times J}$ is a base in $U \otimes_K V$. If I, J are finite sets a natural K-isomorphism $\alpha : (U \otimes_K V)^* \to U^* \otimes V^*$ is then defined using the dual bases

(if $\{e_i\}_{i \in I}$ is a base for U then $\{e_i^*\}_{i \in I}$ where

$$e_i^*(e_k) = \begin{cases} 1 & \text{if } i = k \\ 0 & \text{if } i \neq k \end{cases}$$

is the *dual base* for U^*)

as follows: $\alpha((u_i \otimes v_j)^*) = u_i^* \otimes v_j^*$, by natural extension. One gets the second isomorphism in a similar way.

Ex. 9.10 Using again the result mentioned in the previous exercise, a map $f : K[X] \otimes_K K[Y] \to K[X, Y]$ can be defined (by usual extension)

only on the base $X^i \otimes Y^j \in \mathbb{N} \times \mathbb{N}$. The required K-isomorphism is
the following $f(X^i \otimes Y^j) = X^i Y^j, (i,j) \in \mathbb{N}^2$.

Ex. 9.11 (a) $\mathbb{R}$ is a liniar $\mathbb{Q}$-space of dimension $\aleph$ so that $\mathbb{R} \cong \mathbb{Q}^{(I)}$
as $\mathbb{Q}$-spaces (where $|I| = \aleph$). But in this case $|I \times I| = |I|$ so that the
following $\mathbb{Q}$-isomorphisms hold: $\mathbb{R} \otimes_\mathbb{Q} \mathbb{R} \cong \mathbb{Q}^{(I)} \otimes_\mathbb{Q} \mathbb{Q}^{(I)} \cong$

$\mathbb{Q}^{(I \times I)} \cong \mathbb{Q}^{(I)} \cong \mathbb{R}$.

(b) We use the $\mathbb{Q}$-isomorphism $\mathbb{C} \cong \mathbb{R}^2$ in the following way: $\mathbb{C} \otimes_\mathbb{Q}$
$\mathbb{C} \cong \mathbb{R}^2 \otimes_\mathbb{Q} \mathbb{R}^2 \cong$

$\left(\mathbb{R} \otimes_\mathbb{Q} \mathbb{R} \right)^4 \cong \mathbb{R}^4 \cong \mathbb{C} \times \mathbb{C}$.

(c) $\mathbb{C}$ having as $\mathbb{R}$-space the basis $\{1, i\}$ the $\mathbb{R}$-space $\mathbb{C} \otimes_\mathbb{R} \mathbb{C}$ has
dimension 4. Hence it is isomorph with $\mathbb{R}^4$.

Ex. 9.12 The usual multiplication $u : \mathbb{C} \times \mathbb{C} \to \mathbb{C}$ obviously is bilinear
and balanced and hence induces a ring homomorphism $\mu_\mathbb{C} : \mathbb{C} \otimes_\mathbb{R} \mathbb{C} \to \mathbb{C}$
defined by $\mu_\mathbb{C} \left(\sum_{i=1}^n (z_i \otimes w_i) \right) = \sum_{i=1}^n z_i w_i$. As above, $\mathbb{C}$ having as $\mathbb{R}$-space
the basis $\{1, i\}$ we consider $0 \neq 1 \otimes 1 + i \otimes i$ and $\mu_\mathbb{C} (1 \otimes 1 + i \otimes i) =$
$1 \cdot 1 + i \cdot i = 0$ so that $\ker(\mu_\mathbb{C}) \neq 0$ and hence $\mu_\mathbb{C}$ is not injective.

Ex. 9.13 The formula $(f \otimes g)(x \otimes y) = f(x) \otimes g(y)$ together with the
fact that $\{x \otimes y | x \in M, y \in N\}$ is a generating set for $M \otimes N$ proves
immediately the first assertion.

Obviously, $\ker(f \otimes g)$ includes the subgroup S generated by the two
sets mentioned above. Now, if $p_S : M \otimes N \to (M \otimes N)/S$ is the canonical projection and for $x, x_1 \in M, y, y_1 \in N$ we have $a = x - x_1 \in \ker(f)$
and $b = y - y_1 \in \ker(g)$ then $x_1 \otimes y_1 = x \otimes y + (a \otimes y + x \otimes b + a \otimes b)$
and hence $(a \otimes y + x \otimes b + a \otimes b \in S) \, p_S(x \otimes y) = p_S(x_1 \otimes y_1)$. So, there is
a balanced map $\varphi : M' \times N' \to (M \otimes N)/S$ defined by $\varphi(f(x), g(y)) =$
$p_S(x \otimes y)$ and then, a homomorphism $\psi : M' \otimes N' \to (M \otimes N)/S$ such
that $\psi(x' \otimes y') = \varphi(x', y')$. Then $\psi(f(x) \otimes g(y)) = p_S(x \otimes y)$ and so
$\psi \circ (f \otimes g) = p_S$.

Finally, if $t \in \ker(f \otimes g)$ then $p_S(t) = \psi(f \otimes g)(t) = \psi(0) = 0$
and $t \in S = \ker(p_S)$. Hence $\ker(f \otimes g) = S$.

Ex. 9.14 For $r + I \in R/I$ and $a \in I$, denoting by $\overline{\otimes}$ the tensor in
$I \otimes (R/I)$ we have $\left(1_I \otimes 1_{R/I} \right) (a \overline{\otimes}(r + I)) = a \otimes (r + I) = 1 \otimes (ar + I) =$

$1 \otimes 0 = 0$ and hence $\ker\left(1_I \otimes 1_{R/I}\right) = I \otimes_R (R/I)$. Using the previous exercise, the kernel T of $1_I \otimes p_I : I \otimes R \to I \otimes (R/I)$ is generated by $\{a \otimes r \mid a \in I, r \in R\}$. It is readily checked that the canonical homomorphism $f : I \otimes R \to I, f(a \otimes r) = ar$ maps T on I^2. Since $I^2 \neq I$ clearly $T \neq I \otimes R$ and hence $I \otimes (R/I) \neq 0$. Then $i \otimes 1_{R/I}$ is not injective.

Chapter 10

Artinian and Noetherian Rings

Ex. 10.1 $\mathbb{Z}[i]$ is noetherian together with $\mathbb{Z}$. It is not artinian: $3\mathbb{Z}[i] \supset 3^2\mathbb{Z}[i] \supset .. \supset 3^n\mathbb{Z}[i] \supset ..$ is an infinite decreasing sequence of ideals in $\mathbb{Z}[i]$.

Ex. 10.2 (a) Simple verifications are needed in order to prove that the given sets are left ideals in T. Conversely, let L be a left ideal of T. Consider the following subsets $N = \left\{ (r,x) \in R \oplus M \mid \begin{pmatrix} r & x \\ 0 & 0 \end{pmatrix} \in L \right\}$ and $I = \left\{ s \in S \mid \begin{pmatrix} 0 & 0 \\ 0 & s \end{pmatrix} \in L \right\}$. Again, one verifies that: N is a left submodule of $R \oplus M$, I is a left ideal of S, $M \cdot I \subseteq N$ and $L = \begin{pmatrix} p_R(N) & p_M(N) \\ 0 & I \end{pmatrix}$.

(b) Notice that the opposite ring of T is isomorphic with $\begin{pmatrix} S^{op} & M \\ 0 & R^{op} \end{pmatrix}$ so it suffices to prove only one case, e.g. the left one. Suppose that R, S are left noetherian rings and M is a left noetherian R-module. Using (a) T is also left noetherian. Conversely, if T is left noetherian then R, S are also left noetherian because the maps $f : T \to R, g : T \to S$ defined by $f\begin{pmatrix} r & x \\ 0 & s \end{pmatrix} = r, g\begin{pmatrix} r & x \\ 0 & s \end{pmatrix} = s, r \in R, s \in S, x \in M$ are surjective ring homomorphisms. The R-module

$R \oplus M$ has via f a natural structure of T-module isomorphic with the ideal $\begin{pmatrix} R & M \\ 0 & 0 \end{pmatrix}$ and hence is a left noetherian T-module or even left noetherian R-module. But M is a R-submodule of $R \oplus M$ and hence is a left noetherian R-module.

Ex. 10.3 One applies the previous exercise: $\mathbb{Z}$ and $\mathbb{Q}$ are noetherian rings and $\mathbb{Q}$ is not a noetherian (finitely generated) $\mathbb{Z}$-module (abelian group).

Ex. 10.4 Another application of 10.2: $\mathbb{Q}$ and $\mathbb{R}$ are (trivially as fields) artinian but $\mathbb{R}$ is not an artinian $\mathbb{Q}$-module (linear space).
  Generalization. For $K \subset L$ division rings and $[L : K] = \infty$ the ring $\begin{pmatrix} K & L \\ 0 & L \end{pmatrix}$ is right artinian but not left artinian.

Ex. 10.5 Another application of 10.2.

Ex. 10.6 Another application of 10.2.

Ex. 10.7 If K is a field, $K[X]$ is a principal ideal domain and hence a noetherian ring (one uses a stronger result: a ring R is right noetherian iff each right ideal is finitely generated). Obviously, it is not an artinian ring: $K[X] \supset (X) \supset (X^2) \supset .. \supset (X^n) \supset ..$ is an infinite decreasing sequence of ideals in $K[X]$.

Ex. 10.8 Equivalently, we have to show that each ideal I in R is finitely generated. Let us denote by $I[X] = I \cdot R[X]$. As an ideal in $R[X]$ it is finitely generated, i.e. there are polynomials $f_1, f_2, .., f_n$ that generate $I[X]$ in $R[X]$. If $f_k = \sum_{s=0}^{m_k} a_{ks} X^s, 1 \le k \le n$ then the ideal (finitely) generated by $A = \{a_{ks} | 0 \le s \le m_k; 1 \le k \le n\}$ in $R[X]$ is precisely $I[X]$. A simple judgement concerning the degrees of the polynomials shows that A generates also I in R and hence I is finitely generated.

Ex. 10.9 If $\overline{Y} = Y + (X^2, XY) \in R$ then $(\overline{Y}) \supset (\overline{Y^2}) \supset ..$ is a strictly descending chain of ideals in R so that it is not an artinian ring.

Another method. One shows, $N(R)$ denoting the nilradical (see 13.11), that $R/N(R) \cong \mathbb{Q}[[Y]]$ and as above $(Y) \supset (Y^2) \supset ..$ is a strictly descending chain of ideals in $R/N(R)$. Hence, together with $R/N(R)$, R is not artinian.

Ex. 10.10 Let $I_1 \supset I_2 \supset ..$ be a strictly descending chain of right ideals

of R. Using a natural notation one verifies that $\begin{pmatrix} I_1 & I_1 & \cdots & I_1 \\ 0 & 0 & \cdots & 0 \\ .. & .. & .. & .. \\ 0 & 0 & \cdots & 0 \end{pmatrix} \supset$

$\begin{pmatrix} I_2 & I_2 & \cdots & I_2 \\ 0 & 0 & \cdots & 0 \\ .. & .. & .. & .. \\ 0 & 0 & \cdots & 0 \end{pmatrix} \supset ..$ is a strictly descending chain of right ideals

in $\mathcal{M}_n(R)$. Hence if $\mathcal{M}_n(R)$ is artinian then R has the same property. Conversely, if R is artinian , denote by

$$D_n(R) = \left\{ \begin{pmatrix} r & 0 & \cdots & 0 \\ 0 & r & \cdots & 0 \\ .. & .. & .. & .. \\ 0 & 0 & \cdots & r \end{pmatrix} \middle| r \in R \right\}. \; D_n(R) \text{ is clearly isomorphic}$$

with R and so it is artinian. In the sequel we use the "matric units" E_{ij}^a which are matrices with all the entries zero, except for the element on the i-th row and the j-th column, which is a. If we consider $R_{ij} = \left\langle E_{ij}^r | r \in R \right\rangle$ for each $1 \leq i, j \leq n$ these are artinian as $D_n(R)$-modules (being $D_n(R)$-isomorphic to the $D_n(R)$-module $D_n(R)$). Moreover, clearly $\mathcal{M}_n(R) = \bigoplus_{i,j=1}^{n} R_{ij}$ and hence (direct sums of finitely many artinian (noetherian) modules are artinian (noetherian)) $\mathcal{M}_n(R)$ is an artinian $D_n(R)$-module. Since every right ideal of $\mathcal{M}_n(R)$ is also a $D_n(R)$-module, we finally obtain that $\mathcal{M}_n(R)$ is an artinian ring. The noetherian case is similar.

Ex. 10.11 We prove a stronger result: let $f : R \to R'$ be a surjective ring homomorphism; if $\ker(f)$ and $R' = im(f)$ are left (right) artinian (or noetherian) then R has the same property. In what follows we deal with the artinian case. If $I_1 \supseteq I_2 \supseteq .. \supseteq I_n \supseteq ..$ is a decreasing sequence of right(left) ideals the hypothesis assure there are $m, l \in \mathbb{N}^*$

such that $I_m \cap \ker(f) = I_{m+1} \cap \ker(f) = ..$ resp. $f(I_l) = f(I_{l+1}) = ..$
For $k = \max(m, l)$ one shows that $I_k = I_{k+1} = ..$ and hence R is left
(right) artinian.

Now, if $R_1, R_2, .., R_n$ are left (right) artinian (or noetherian) and
$R = R_1 \times R_2 \times .. \times R_{n-1}$ we apply the previous result to $f = p_n : R \to R_n$
the canonical projection in order to infer that R has the same property.
The rest is a simple induction.

Ex. 10.12 Let $0 \neq a \in R$ be an artinian ring. Consider the decreasing
sequence of ideals $Ra \supseteq Ra^2 \supseteq .. \supseteq Ra^n \supseteq ..$ The ring R being artinian,
for a suitable $m \in \mathbb{N}^*$ we have $Ra^m = Ra^{m+1} = ..$ Hence $a^m = a^{m+1}r$,
for an element $r \in R$ and so $1 = ar$, R being an integral domain, i.e. a
is a unit. The converse is obvious.

Ex. 10.13 Let $H_1 \supseteq H_2 \supseteq .. \supseteq H_n \supseteq ..$ be a decreasing sequence of
right ideals in the ring R. Then $l(H_1) \subseteq l(H_2) \subseteq .. \subseteq l(H_n) \subseteq ..$ is
an increasing sequence of left ideals. The ring R being left noetherian,
for a suitable $m \in \mathbb{N}^*$ we have $l(H_m) = l(H_{m+1}) = ..$ and hence
$rl(H_m) = rl(H_{m+1}) = ..$ Applying an annihilator condition we get
$H_m = H_{m+1} = ..$

Ex. 10.14 We have already seen one implication in the previous exercise. In what follows we treat another one, say, each left noetherian
ring R which satisfies the annihilator conditions is also right noetherian.
First, as above, one verifies (as above) that R is right artinian. Then,
the opposite ring R^{op} is left artinian, and one uses a well-known result
of Hopkins - each left artinian ring is also a left noetherian ring. If R^{op}
is left noetherian finally R is right noetherian. The other implications
follow similarly.

Ex. 10.15 Let $P \in \mathcal{NF}$ be a maximal element. If P would not be a
prime ideal, there should exist elements $r, s \in R$ such that $ab \in P, a \notin$
$P, b \notin P$. Then $P + Rr$ and $P + Rs$ (which properly contain P) must
be finitely generated. Observe that the R-module
$$(P + Rs) / (P + Rr)(P + Rs)$$ is finitely generated because its R-
module structure is the same with its $R/(P + Rr)$-module structure
and $R/(P + Rr)$ is noetherian (the astute reader will notice that P being maximal in $\mathcal{NF}$, R/P is noetherian). Hence $P/(P + Rr)(P + Rs)$

is a finitely generated R-module and, together with $P+Rr$ and $P+Rs$, P would be a finitely generated R-module, a contradiction.

Finally, let us verify that $(\mathcal{NF}, \subseteq)$ is inductive (each chain has an upper bound). If $\{I_j\}_{j \in J}$ is a chain of non-finitely generated ideals in R then surely $U = \bigcup_{j \in J} I_j$ is an ideal in R. The ideal U cannot be finitely generated; otherwise, for each $j \in J$ such that I_j includes all the generators of U we would have $U = I_j$ finitely generated, a contradiction.

Ex. 10.16 In a noetherian ring each ideal is finitely generated. Conversely, if each prime ideal in a ring R is finitely generated, using the previous exercise, in R the set of all the non-finitely generated ideals is empty. Hence R is noetherian.

Ex. 10.17 Consider the $R/(U \circ V)$-modules exact sequence
$$0 \to U/(U \circ V) \to R/(U \circ V) \to R/U \to 0.$$
Being trivially (the ring R/U is artinian) an artinian R/U-module, R/U is also an artinian $R/(U \circ V)$-module. Further, $U/(U \circ V)$ is also an artinian $R/(U \circ V)$-module, because (the ring R/V is artinian) it is a finitely generated R/V-module. Hence, $R/(U \circ V)$ must be (extentions of artinian modules by artinian modules are artinian) also an artinian $R/(U \circ V)$-module and so the corresponding ring is artinian. The noetherian case is treated similarly.

Ex. 10.18 If P is a prime ideal, R/P is an integral domain, artinian together with R. One has now to use 10.12 in order to see that R/P is a field. Hence P is a maximal ideal.

Ex. 10.19 Each artinian commutative ring is also noetherian (Theorem of Hopkins, see above) so that the proof of the direct implication is completed by the previous exercise. Conversely, let R be a noetherian ring such that each prime ideal is maximal and suppose $R \cong R/(0)$ is not artinian. There is an ideal U of R maximal (the corresponding set is not empty) relative to the property that R/U is not artinian. U cannot be prime by our hypothesis: it would be also maximal and R/U would be trivially artinian being a field. Hence, there are ideals I, J in R such that $I \circ J \subseteq U$ and $U \subset I, U \subset J$. By the maximality of U, the quotient rings $R/I, R/J$ are artinian. Moreover, R being noetherian, J is finitely generated and hence using 10.17, $R/I \circ J$ and R/U is artinian, a contradiction.

Ex. 10.20 Each nilpotent left ideal is obviously nil. The converse reduces to 17.12.

Ex. 10.21 The solution is an immediate consequence of the two following results: *in a right noetherian ring the prime radical is the largest nilpotent right ideal*, and, *the largest nil left ideal*. The reader can find these results in [17],p.69-70.

Ex. 10.22 (a) Using the theorem of Hopkins R is also left noetherian. The set of all the nilpotent left ideals is not empty (e.g. (0)) so one simply uses Zorn's lemma.

We show that $N = \mathcal{R}(R)$ (see 13.11), the prime radical of R, which solves (b) and (c).

First, if I is a nilpotent left ideal of R then there is a $n \in \mathbb{N}^*$ such that $I^n = (0) \subseteq \mathcal{R}(R) = \cap\{P|P$ prime ideal$\}$ and hence $I \subseteq \mathcal{R}(R)$. Next, in a similar way to the solution of 13.11, one shows that in a left artinian ring R the prime radical $\mathcal{R}(R)$ is a nilideal. Finally, an analogous result of the previous exercise is needed: *any left nilideal in a left artinian ring is nilpotent*, in order to prove that $\mathcal{R}(R)$ is nilpotent.

(d) Using 10.12 we obtain $\mathcal{R}(R) = \cap\{P|P$ prime ideal$\} =$
$\{M|M$ maximal ideal$\} = rad(R)$.

Chapter 11

Socle and Radical

Ex. 11.1 If R is a division ring (this is also true for semisimple rings, see next chapter) surely $s(R) = R \neq 0$.

Conversely, first observe that if A is a right minimal ideal and $0 \neq a \in A$ then $A = Ra$ (clearly $0 \neq Ra \subseteq A$). Now, if R is a domain with $0 \neq Ra$ minimal right ideal then for each $r \in R$ we have $Rra = Ra$ (as above $0 \neq Rra \subseteq Ra$) and then $a \in Rra$. If $s \in R$ is such that $a = sra$ then $a(1 - sr) = 0$ and $1 = sr$ (because $a \neq 0$ and R is a domain). Similarly, one gets a right inverse for r.

Ex. 11.2 For $n \in \{0,1\}$ clearly the socle is 0. Let $n = p_1^{r_1}..p_k^{r_k}$ the prime number decomposition of $n > 1$. First observe that $\mathbb{Z}/m\mathbb{Z}$ is a simple $\mathbb{Z}$-module (abelian group) iff m is a prime number (indeed, if m is a prime number then $\mathbb{Z}/m\mathbb{Z}$ is a field and a simple $\mathbb{Z}$-module; conversely if q is a proper divisor of m then $q\mathbb{Z}/m\mathbb{Z}$ is a proper $\mathbb{Z}$-submodule of $\mathbb{Z}/m\mathbb{Z}$). Using obvious isomorphisms $\frac{n}{p_i}\mathbb{Z}/n\mathbb{Z} \cong \mathbb{Z}/p_i\mathbb{Z}$ these are simple submodules of $\mathbb{Z}/n\mathbb{Z}$ and $\sum_{i=1}^{k} \frac{n}{p_i}\mathbb{Z}/n\mathbb{Z}$

$$= \left(\sum_{i=1}^{k} \frac{n}{p_i}\mathbb{Z}\right)/n\mathbb{Z} = \frac{n}{p_1..p_k}\mathbb{Z}/n\mathbb{Z} \subseteq s(\mathbb{Z}_n).$$

Conversely, if $q\mathbb{Z}/n\mathbb{Z}$ with $n = qm$ is a simple submodule of $\mathbb{Z}/n\mathbb{Z}$, according to the isomorphism $q\mathbb{Z}/n\mathbb{Z} \cong \mathbb{Z}/m\mathbb{Z}$, m must be one of the prime numbers $p_1, .., p_k$ and so $q = \frac{n}{p_i}$ for a suitable $i \in \{1, .., k\}$. Hence $s(\mathbb{Z}_n) = \frac{n}{p_1..p_k}\mathbb{Z}_n$.

Ex. 11.3 If I is essential in K then for each right ideal A included in K, $A \neq 0 \Rightarrow I \cap A \neq 0$ and this holds also for any right ideal included in J (because $J \subset K$). Moreover, $I \cap A \subseteq J \cap A$ and hence $J \cap A \neq 0$; so J is also essential in K. Conversely, with the above notations $A \neq 0 \Rightarrow J \cap A \neq 0$ because J is essential in K and then $I \cap A = I \cap (J \cap A) \neq 0$ because I is essential in J.

Ex. 11.4 Let $A \neq 0$ be a right ideal in R. We distinguish two cases: if $f(A) = 0$ then $A \subseteq \ker(f) = f^{-1}(0) \subseteq f^{-1}(J)$ so that obviously $0 \neq A = A \cap f^{-1}(J)$. If $f(A) \neq 0$ then $f(A) \cap J \neq 0$ (J being essential in R'). Hence there is an element $0 \neq a \in A : 0 \neq f(a) \in J$ so that $a \in A \cap f^{-1}(J) \neq 0$.

Another solution. Using 4.13 and 2.6 clearly $f^{-1}(f(A) \cap J) = f^{-1}(f(A)) \cap f^{-1}(J) = (A + \ker(f)) \cap f^{-1}(J) = \ker(f) + (A \cap f^{-1}(J))$. Now, if $A \cap f^{-1}(J) = 0$ we deduce $f^{-1}(f(A) \cap J) = \ker(f)$. Surely $f(A) \cap J = f(A')$ for a suitable $A' \subseteq A$ and so $f^{-1}(f(A')) = \ker(f)$ implies $A' + \ker(f) = \ker(f)$ and $A' \subseteq \ker(f)$. Hence $f(A) \cap J = f(A') = 0$.

Ex. 11.5 First, we show that each simple ideal S is contained in each essential one, E. From $S \neq 0$ we have $E \cap S \neq 0$ and hence $E \cap S = S$ (because surely $S \cap E \subseteq S$). Then $S \subseteq E$ and so $s(R) = \sum \{S | S$ simple ideal in $R\} \subseteq I = \cap \{E | E$ essential ideal in $R\}$. Conversely, it suffices to verify that the above intersection I is semisimple (i.e. sum of simple ideals). We verify that each ideal A in I is a direct summand. Indeed, let B be an ideal of R maximal relative to the property $A \cap B = 0$. Then $A \oplus B$ is essential in R and then $A \subseteq I \subseteq A \oplus B$. Using 2.6 we obtain $I = I \cap (A \oplus B) = A \oplus (B \cap I)$ and hence A is a direct summand of I.

Ex. 11.6 If I is a superfluous ideal in R let M be a maximal ideal of R. Then surely $M \subseteq I + M$ and so $I + M \in \{M, R\}$. But $I + M = R$ implies $M = R$, impossible (because I is superfluous) so that $I + M = M$ and hence $I \subseteq M$. From this follows $I \subseteq rad(R) = \cap \{M | M$ maximal ideal in $R\}$.

Ex. 11.7 If $x \in I$ we show that xR is superfluous. Indeed, if for a right ideal A, $xR + A = R$ holds then $1 = xr + a$ for suitable elements

$r \in R$ and $a \in A$. Since $a = 1 - xr \in 1 + I$, a must be a unit in R and hence $A = R$. Then $I \subseteq rad(R)$ using the previous exercise.

Ex. 11.8 First, it is readily checked that if I is a finitely generated right ideal then I is *compact*, i.e. each cover $I \leq \sum_{i \in I} B_i$ (B_i right ideals) has a finite subcover $I \leq \sum_{i \in F} B_i$ (for a suitable finite subset F of I).

Next, suppose that a finitely generated right ideal I is not superfluous. Then there is a right ideal $A \neq R$ such that $I + A = R$ and hence $I \not\subseteq A$. The set $\mathcal{A} = \{B$ right ideal in $R | I \not\subseteq B, I + B = R\}$ is inductive (one uses the compactness of I) and so, by Zorn's lemma, let M be maximal in $\mathcal{A}$. Moreover, M is maximal in R (indeed, if $M \subset N$ then $I \subseteq N$ which implies $R = I + M \subseteq N$) and hence $I \not\subseteq rad(R)$ (otherwise $I \subseteq rad(R) \subseteq M$ and $M \notin \mathcal{A}$).

Ex. 11.9 From 11.6 one has $\sum \{I | I$ superfluous ideal in $R\} \subseteq rad(R)$ $= \bigcap \{M | M$ maximal ideal in $R\}$. Conversely, let $x \in rad(R)$. We prove that Rx is a superfluous ideal in R. Suppose that $Rx + A = R$. If $A \neq R$ then $x \notin A$ and, using Zorn's lemma, let M be maximal relative to the properties $A \subseteq M$ and $x \notin M$. Since $M + Rx = R$ the ideal M is maximal in R and hence $x \in rad(R) \subseteq M$, a contradiction. So $A = R$ and Rx is superfluous and $rad(R) \subseteq \sum \{I | I$ superfluous ideal in $R\}$.

Ex. 11.10 Let $n = p_1^{r_1}..p_k^{r_k}$ the prime number decomposition of n with $r_i \in \mathbb{N}^*, 1 \leq i \leq k$. The maximal (or prime) ideals in $\mathbb{Z}$ are the principal ideals generated by the prime numbers. The maximal ideals which include $n\mathbb{Z}$ are exactly $\{p_i\mathbb{Z} | 1 \leq i \leq k\}$. Moreover, $\bigcap_{i=1}^{k} p_i\mathbb{Z} = p_1..p_k\mathbb{Z}$ and hence $rad(\mathbb{Z}_n) = \left(\bigcap_{i=1}^{k} p_i\mathbb{Z}\right) /n\mathbb{Z} = p_1..p_k\mathbb{Z}/n\mathbb{Z}$.

Ex. 11.11 If $rad(R) = \bigcap \{M | M$ maximal modular right ideal of $R\}$ (see also 2.4 and 1.2) let $H = \{r \in R | xry$ is quasi-regular in $R\}$.

For $rad(R) \subseteq H$ we first observe that if an element $a \in R$ is not right quasi-regular then there is a modular maximal ideal M such that $a \notin M$. Indeed, if a is not right quasi-regular, the right ideal

$U = \{ar - r | r \in R\}$ does not contain a. The Zorn's lemma shows the existence of a right ideal M maximal in the set of all the right ideals which contain U but not the element a. One verifies that this is a maximal right ideal in R. Moreover, M is modular since $ar - r \in M$ for each $r \in R$. Next, if $b \notin H$ there are elements $x, y \in R$ such that xby is not right quasi-regular. Then, there is a maximal modular right ideal which does not contain xby and hence, does not contain b, and $b \notin rad(R)$.

Conversely, one has to prove first that if M is a maximal modular right ideal in R then $H \subseteq R : M$ ([18] Lemma 6.17; see also 2.13). Since $rad(R) \neq R$ there actually exist maximal modular right ideals M in R and $H \subseteq \cap \{R : M | M$ maximal modular right ideal in $R\}$. Hence $R.H \subseteq M$ for each M and M being modular, we also have $H \subseteq M$. Finally, $H \subseteq rad(R)$.

Ex. 11.12 First, we observe that in a ring with identity R an element $r \in R$ is right quasi-regular iff $1 - r$ is right invertible (indeed, $r + s - rs = 0 \Leftrightarrow (1 - r)(1 - s) = 1$). Next, we show that

$M = \{r \in R | xry$ is right quasi-regular in $R, \forall x \in R\} =$

$N = \{r \in R | xr$ is right quasi-regular in $R, \forall x \in R\}$. The ring having identity, clearly $M \subseteq N$. Conversely, if $xr + t - xrt = 0$ then $xry + ty - xrty = (xr + t - xrt)y = 0$ and so $N \subseteq M$. Finally, our result follows from the previous exercise.

Ex. 11.13 We use 11.11. Let $A = (a_{ij}) \in rad(\mathcal{M}_n(R))$ and $x, y \in R$. We verify that $xa_{ij}y$ is a right quasi-regular element. Consider $B = E_{1i}^x A E_{j1}^y = E_{11}^{xa_{ij}y}$ (as before we denote by E_{ij}^a a matrix with zero entries excepting the $i - j$ one which is a) which belongs to $rad(\mathcal{M}_n(R))$ together with A. Then there is a matrix $C = (c_{ij})$ such that $B + C - BC = 0$ and hence $xa_{ij}y + c_{11} - xa_{ij}yc_{11} = 0$.

Conversely, if for every matrix $A \in \mathcal{M}_n(R)$ one clearly has $A = \sum_{i,j} E_{ij}^{a_{ij}}$ so, if $A \in \mathcal{M}_n(rad(R))$ it suffices to prove that any $E_{ij}^{a_{ij}} \in \mathcal{M}_n(rad(R))$. First, for an arbitrary matrix $Y = (y_{ij})$, we prove that $E_{ij}^{a_{ij}}Y$ is right quasi-regular. Together with a_{ij} all the products $a_{ij}y_{kl}$ are right quasi-regular and hence there are elements z_{ik} such that $a_{ij}y_{jk} + z_{ik} - a_{ij}y_{jk}z_{ik} = 0$. Denoting by $Z = \sum_{k=1}^{n} E_{ik}^{z_{ik}}$ from the previous equalities follows $E_{ij}^{a_{ij}}Y + Z - E_{ij}^{a_{ij}}YZ = 0$. The same thing can be done for $XE_{ij}^{a_{ij}}Y$ for arbitrary $X, Y \in \mathcal{M}_n(R)$ and so $E_{ij}^{a_{ij}} \in \mathcal{M}_n(rad(R))$.

Ex. 11.14 According to 11.11 let $e \in R$ a right quasi-regular idempotent element. If $e \circ e' = 0$ then $e = e + ee' - ee' = e(e + e' - ee') = e(e \circ e') = e \cdot 0 = 0$.

$\mathcal{M}_n(K)$ is a simple ring (see 2.20)and so $rad(\mathcal{M}_n(K)) = 0$. $rad(B)$ is also zero according to the first part of the exercise.

Ex. 11.15 For commutativity, denoting by $A(a,b,c) = \begin{pmatrix} a & 0 & 0 \\ b & a & 0 \\ c & 0 & a \end{pmatrix}$ one verifies $A(a, b, c) \cdot A(x, y, z) = A(ax, ay + bx, az + cx) = A(x, y, z) \cdot A(a, b, c)$. Further we show that $rad(V) = \{A(a, b, c) | a = 0\}$. K being a field, V is a commutative ring with identity and hence we use 11.12: $I_3 - A(a, b, c) \cdot A(x, y, z)$ is invertible for each $x, y, z \in K$ iff this matrix has a non-zero determinant. But this leeds to $(1 - ax)^3 \neq 0$ for every $x \in K$ and hence $a = 0$ (indeed, if $a \neq 0$ the $(1 - aa^{-1})^3 = 0$).

Ex. 11.16 If $\hat{\mathbb{Z}}_{(p)}$ denotes the ring (with identity) of the p-adic numbers, i.e. $\{a = a_0 + a_1 p + .. + a_n p^n + .. | 0 \leq a_n < p, a_n \in \mathbb{Z}, n \in \mathbb{N}^*\}$

together with the addition and multiplication according to the "remainder to the division by p" then we prove that $rad(\hat{\mathbb{Z}}_{(p)}) = J = \{a | a_0 = 0\}$. Indeed, it is known that (with the above notations) a p-adic integer a is a unit iff $a_0 \neq 0$. So J is the set of all the nonunits in $\hat{\mathbb{Z}}_{(p)}$ and (by straightforward computations) actually is an ideal. Hence (see 13.23) $\hat{\mathbb{Z}}_{(p)}$ is a local ring and $rad(\hat{\mathbb{Z}}_{(p)}) = J$.

Ex. 11.17 We use the following result (see 11.12):

$rad(R) = \{r \in R | \forall s \in R$ the element $1 - rs$ is right invertible$\}$ in order to prove that $rad(T)$ consists of all the triangular matrices that have zero entries on the diagonal. We search for the matrices $A = (a_{ij}) \in T$ such that $I_n - AB$ is right invertible for each $B \in T$. The field K being commutative a matrix is right invertible iff it is left invertible (and hence invertible). Now, $I_n - AB$ is invertible iff $det(I_n - AB) = (1 - a_{11}b_{11})(1 - a_{22}b_{22})..(1 - a_{nn}b_{nn}) \neq 0$ for each $B = (b_{ij}) \in T$ and hence iff $a_{11} = a_{22} = .. = a_{nn} = 0$.

In $T/rad(T)$ we have $A + rad(T) = B + rad(T)$ iff $A - B \in rad(T)$, that is if the matrices have the same diagonal (see also 2.22). Hence, in $T/rad(T)$ we can choose as representatives only diagonal matrices (i.e.$i \neq j \Rightarrow a_{ij} = 0$) and deduce immediately the commutativity of $T/rad(T)$ from the commutativity of K (diagonal matrices multiply directly on the diagonal).

Remark. If $D_n(K)$ denotes the subring of all the diagonal matrices of $\mathcal{M}_n(K)$ then obviously $T/rad(T) \cong D_n(K)$.

Ex. 11.18 Acording to 11.11 it suffices to observe that $f(a \circ b) = f(a + b - ab) = f(a) + f(b) - f(a)f(b) = f(a) \circ f(b)$ and hence right quasi-regular elements (resp. right ideals) are preserved by ring homomorphisms.

Ex. 11.19 One has to use 14.15. Surely $\mathcal{N}(R[X]) \subseteq Rad(R[X])$. Conversely, let $f = a_0 + a_1X + .. + a_nX^n \in Rad(R[X])$. So (see 11.12) $1 + Xf = 1 + a_0X + ..a_nX^{n+1}$ is a unit and hence $a_0, a_1, .., a_n$ are nilpotent. But then f is nilpotent in $R[X]$ and $f \in \mathcal{N}(R[X])$.

Ex. 11.20 If $r \in R$ according to 11.12 $1 + rX$ is a unit and hence (see 14.15) r is nilpotent.

Ex. 11.21 Results of S.A.Amitsur show that more generally
    (see [17],p.59,ex.14) if $N = R \cap rad(R[X])$ then $rad(R[X]) = N[X]$.
    First, one shows that the radical $R[X]$ is invariant under all the automorphisms of $R[X]$.
    Next, if $rad(R[X]) \neq 0$ consider $f(X) \in R[X]$ a polynomial of minimal degree in $rad(R[X])$. According to 8.9 $X \mapsto X + 1$ is an automorphism of $R[X]$ and then $f(X + 1) \in rad(R[X])$ and, by the minimality $f(X + 1) - f(X) = 0$. Hence, if $char(R) = 0$ then only $f(X) = a \in R$ is possible, with $a \neq 0$.

Ex. 11.22 We observe that the ring $K \langle X \rangle$ of power series contains a single maximal right ideal, namely the ideal (X) which consists of all power series of the form $f = \sum_{i=1}^{\infty} a_iX^i (a_i \in K)$. In fact, more can be proven: the right (left) ideals of $K \langle X \rangle$ are exactly $(0), (X^0) =$

$K\langle X\rangle, (X), (X^2), .., (X^n), ..$ (indeed, if I is a right ideal and X^k is a smallest power among the powers of X occurring in the representations of the non-zero elements of I then one shows that $I = \left(X^k\right)$; see [16],p.38). Since $K\langle X\rangle$ has identity, this is also the only modular maximal right ideal in $K\langle X\rangle$ and so $rad(K\langle X\rangle) = (X)$.

Chapter 12

Semisimple Rings

Ex. 12.1 If K is any field, an infinite product K^I is a nonsemisimple ring. But it is the product of the (nonisomorphic) simple right ideals $\{P_i\}_{i \in I}$ where $P_i = im(q_i)$, and $q_i : K \to K^I$ are the canonical injections. Indeed, it suffices to remark that isomorphic rings (or even modules) have the same annihilator ideal.

Ex. 12.2 If R is semisimple, every proper right ideal in R is a direct summand of R and can not be essential. Conversely, let I be an arbitrary right ideal of R. Using Zorn's lemma one can verify that there is another right ideal J maximal relative to the property $I \cap J = 0$ (a so called *complement* of I). Then $I + J$ is an essential right ideal (indeed, if $(I + J) \cap K = 0$ then using the modularity and $I \cap (J+K) \subseteq (I+J) \cap (J+K) \overset{mod}{=\!=} ((I+J) \cap K) + J = J$ one obtains $I \cap (J+K) = 0$ and hence $K = 0$ by the maximality of J) and by hypothesis $I \oplus J = R$. Hence every right ideal of R is a direct summand and R is semisimple.

Ex. 12.3 If R is semisimple each ideal I is a direct summand of R. Now if $R = I \oplus J$ then surely $R/I \cong J$. But if each ideal in R is a direct summand an ideal I has the same property: indeed, if ι is an ideal in I and $R = \iota \oplus U$ then using 2.6 $I = \iota \oplus (I \cap U)$. Hence each ideal and each quotient ring are semisimple.

Ex. 12.4 No. Obviously $\mathbb{Q}$ is simple and hence semisimple as a field.

In $\mathbb{Z}$ no proper ideal is a direct summand so $\mathbb{Z}$ is a non-semisimple subring of $\mathbb{Q}$. One can easily generalize this example.

Ex. 12.5 The ring R' has a natural structure of R-module via f (unital, being a surjective ring homomorphism) as follows: $r \cdot r' \stackrel{def}{=} f(r).r'$ The ring homomorphism being surjective, each ideal from R' is a R-submodule in $_R R'$. Hence R' is semisimple iff $_R R'$ is semisimple (each ideal in R' is a direct summand $\Leftrightarrow$ each R-submodule of $_R R'$ is a direct summand).

Ex. 12.6 $\mathbb{Z}_n$ is semisimple iff $s(\mathbb{Z}_n) = \mathbb{Z}_n$ and one has only to use 11.2: $\mathbb{Z}_n$ is semisimple iff n is square-free.

Ex. 12.7 If each R_i is semisimple, it is a finite direct sum of simple ideals and a finite direct product (or sum) has the same structure. Hence if I if finite, $\prod_{i \in I} R_i$ is semisimple too. The converse follows similarly to our first exercise 12.1.

Ex. 12.8 No. We may construct the chain of total matrix algebras $\{k_n\}, n = 2^m, m \in \mathbb{N}^*$, i.e. $k_2 \subset k_4 \subset k_8 \subset ..$ The union is a regular ring which does not satisfy the ascending chain condition for direct summands and hence according to 12.14 (in a semisimple ring each (left) ideal is a direct summand), is not semisimple.

Ex. 12.9 In fact, more can be proven: *if R is a ring with identity, each ideal which is a direct summand is generated by a central idempotent.* Indeed, if $R = A \oplus B$ is a direct decomposition in (left) ideals, let $1 = e + b(e \in A, b \in B)$. Then $e = e^2 + eb$ and $eb = e - e^2 \in A \cap B = \{0\}$ so e is an idempotent element. Further, if $a \in A$ then $a = ae + ab$ and hence $ab = a - ae \in A \cap B = \{0\}$ and $a = ae$ or $A = Ae$. Symmetrically, $A = eA$ holds too. Finally, if $r \in R$ then if A, B are ideals one has $re + rb = r.1 = 1.r = er + br$ and so by uniqueness $re = er$; hence e belongs to the center of R.

Ex. 12.10 The ring R being semisimple, Re is a direct summand of R. So Re has an identity, namely e, and hence $eRe = Re$. Now, if Re is a minimal left ideal in R and $0 \neq a \in eRe$ then $Ra = Re$ and hence there is an element $b \in R$ such that $ba = e$. But then $be \in Re = eRe$ is a left inverse for a (don't forget that e is an identity in Re !). Conversely, if $eRe = Re$ is a division ring for each $0 \neq c \in Re$ there is an element $d \in Re$ such that $dc = e$. Hence $Rc = Re$ (one checks the two inclusions) and this is a minimal left ideal.

Ex. 12.11 It suffices to prove one implication, say R left semisimple $\Rightarrow R$ right semisimple, the converse being symmetric.

Let $R = \bigoplus_{i=1}^{n} L_i = \bigoplus_{i=1}^{n} Re_i$ be a decomposition of R in simple left ideals, that is, $\{e_i | 1 \leq i \leq n\}$ is a set of central orthogonal idempotents (i.e. $e_i \neq 0, e_i e_j = \delta_{ij} e_i$) such that $1 = \sum_{i=1}^{n} e_i$. Then clearly $R = \bigoplus_{i=1}^{n} e_i R$ holds and the only thing left to be verified is that $e_i R$ is a simple (right) ideal. Let $0 \neq a \in eR$ (where e denotes one of the e_i). If $a = er$ then $ea = e^2 r = er = a$ and hence $aR \subseteq eR$ so that we only have to verify $aR = eR$. From $ea \neq 0$ and Re simple follows that the map $f : Re \to Ra, f(re) = rea = ra, \forall r \in R$ is a left R-module isomorphism (only the injectivity of f needs care). If now $R = Ra \oplus I$ then the map $g : R \to R, g(ra + x) = re, \forall r \in R, \forall x \in I$ is a ring endomorphism of R and hence g must be (see for example [14], 3.7) a right translation (multiplication) $g = t_b$, with a suitable $b \in R$. But then $e = g(a) = ab$ implies $e \in aR$ or $eR \subseteq aR$ and hence $aR = eR$.

Ex. 12.12 If R is a left semisimple ring then $rad(R) = 0$ (one can show - using also 12.11 - that there are no superfluous ideals and use 11.9). Moreover, if $R = \sum_{i \in I} m_i$ is a sum of minimal left ideals, using $1 \in R$ one deduces $R = \sum_{k=1}^{n} m_{i_k}$ and, using Zorn's lemma, we may suppose that the last sum is direct. Each minimal left ideal being obviously left artinian, this finite direct sum of left artinian ideals is also left artinian. Conversely, if R is left artinian and $rad(R) = 0$ there are only a finite number of maximal ideals $M_i, 1 \leq i \leq s$ such that $\bigcap_{i=1}^{s} M_i = 0$. Hence if we consider the canonical left R-module projections $p_i : R \to R/M_i, 1 \leq i \leq s$ and $f : R \to \bigoplus_{i=1}^{s}(R/M_i)$ (by universality) then $\ker(f) = \bigcap_{i=1}^{s} \ker(p_i) = \bigcap_{i=1}^{s} M_i = 0$ so that R is R-isomorphic with

a submodule of $\bigoplus_{i=1}^{s}(R/M_i)$, a semisimple R-module. But then R is left semisimple.

Ex. 12.13 Follows immediately from the previous exercise using 11.13 and 10.10.

Ex. 12.14 (a)$\Rightarrow$(b) If R is semisimple, it is a finite sum of simple ideals (indeed, if $R = \bigoplus_{i\in I} S_i$ then from $1 \in R$ we see that I has to be finite) and hence an artinian ring. Using 17.14 R is also regular.

   (b)$\Rightarrow$(c) This follows from a well-known result of Hopkins: each left artinian ring is also left noetherian.

   (c)$\Rightarrow$(a) Using again 17.14 (iii), R has only principal ideals. Then, again by this exercise (ii), each ideal of R is a direct summand. Hence R is semisimple.

Ex. 12.15 We know from 1.23 that R is a commutative ring (with identity) all its elements being idempotent. Such a ring is clearly regular (for every $a \in R$: $a = aaa$) so that the first three condition are equivalent by the previous exercise. Finally, each principal ideal (being generated by an idempotent element) has to be a direct summand. If R is not finite for each $0 \neq a \in R$ at least one from Ra and $R(1 - a)$ is not finite and one can easily construct infinite strictly descending (or ascending) chains of ideals (R is not artinian or not noetherian). The converse is similar.

Ex. 12.16 Let $_RM$ be a semisimple R-module and $f : M \to M$ a R-endomorphism. Each submodule being a direct summand, there are submodules S, T such that $M = \ker(f) \oplus S = im(f) \oplus T$. Using a well-known isomorphism theorem, clearly f induces an isomorphism $g : S \to im(f)$ with an inverse, say $h : im(f) \to S$. This inverse h may be extended to a R-endomorphism $k : M \to M$, defining $k(T) = 0$. But now we can verify that $f = fkf$ (indeed, if $x = u + v, u \in \ker(f), v \in S$ then $fkf(x) = fkf(v) = fkg(v) = fhg(v) = f(v) = f(x)$) and hence $End_R(M)$ is regular.

Ex. 12.17 For a ring homomorphism f surely $f(a \circ b) = f(a) \circ f(b)$, the circle composition (see 1.2). The inclusion $f(rad(R)) \subseteq rad(f(R))$ is clear using 11.11. Conversely, if $R/rad(R)$ is semisimple then each factor ring of it is also semisimple (see 12.3) and has obviously zero radical. Now, consider the canonical composition

$R \xrightarrow{\bar{f}} f(R) \xrightarrow{p'} f(R)/f(rad(R))$ with $\bar{f}(r) = f(r), \forall r \in R$ and p' the canonical projection. This composition is surjective (together with $\bar{f}$ and p') and $rad(R) \subseteq \ker(p' \circ \bar{f})$. Hence the composition factors through $R/rad(R)$: there is a surjective (together with the previous composition) ring homomorphism $u : R/rad(R) \to f(R)/f(radR))$ such that $p' \circ \bar{f} = u \circ p$ where $p : R \to R/rad(R)$ is the canonical projection. Then $f(R)/f(rad(R))$ is semisimple too and according to 12.12, $rad(f(R)/f(rad(R))) = 0$ and we obtain the required equality.

Ex. 12.18 We first prove that *each non-nilpotent minimal right (or left) ideal $\underline{m}$ of an arbitrary ring R contains a non-zero idempotent element e and $\underline{m} = Re = eR$*. Indeed, from $\underline{m}^2 \neq \{0\}$ there is an element $a \in \underline{m}$ such that $\{0\} \neq a\underline{m} \subseteq \underline{m}$. This being a minimal right ideal we must have $a\underline{m} = \underline{m}$ and hence there is an element $0 \neq e \in \underline{m}$ such that $ae = a$. This implies $ae^2 = ae = a$ or $a(e^2 - e) = 0$. Consider the right ideal in R defined as $I = \{r \in \underline{m} | ar = 0\}$. From $\{0\} \neq a\underline{m}$ we infer $I \neq \underline{m}$. The latter being minimal we must have $I = \{0\}$ and hence $e^2 - e = 0$. Finally, $\{0\} \neq eR \subseteq \underline{m}$ implies, again by minimality that $eR = \underline{m}$.

Now, as for the original exercise, first observe that it suffices to prove that each element $0 \neq r \in R$ is contained in a sum of minimal right ideals. According to the descending condition from the hypothesis there is a minimal right ideal $\underline{m}_1$ included in (r). The ring having no nonzero nilpotent ideals we must have $\underline{m}_1^2 = \underline{m}_1$ (the other alternative would be $\underline{m}_1^2 = \{0\}$ by minimality) and so, by our previous result, there exists an idempotent element e such that $\underline{m}_1 = eR = (e)$. Next, we consider $a_1 = a - ea$ such that clearly $(a) = \underline{m}_1 \oplus (a_1)$ and one can continue this procedure until $(a) = \underline{m}_1 \oplus \underline{m}_2 \oplus .. \oplus \underline{m}_n \oplus (a_n)$. We have constructed in this way a strictly descending chain $(a) \supset (a_1) \supset (a_2) \supset .. \supset (a_n)$ of principal right ideals. Using again our hypothesis, this chain terminates for a suitable $n \in \mathbb{N}^*$ and $a \in (a) = \underline{m}_1 \oplus \underline{m}_2 \oplus .. \oplus \underline{m}_n$ a finite sum of minimal right ideals (i.e. $R = s(R)$ the socle).

Ex. 12.19 (a) Straightforward verifications show that R is a ring (commutative with identity) and surely a subring with identity of $\mathbb{Q}$ generated by $\left\{\frac{1}{5}, \frac{1}{7}, ..., \frac{1}{p}, ..|p \geq 5 \text{ and prime number}\right\}$ i.e. $R = A\left(\{5, 7, ..\}\right)$ (see also the solution of 1.30). Clearly, $R/2R \cong \mathbb{Z}_2$ and $R/3R \cong \mathbb{Z}_3$ are fields and so the corresponding (principal) ideals are maximal. There are no other maximal ideals: indeed, each $x \notin 2R \cup 3R$ is a unit in R and so the proper ideals I of R satisfy $I \subseteq 2R \cup 3R$. Moreover, one verifies that $I\backslash 2R \neq \emptyset$ and $I\backslash 3R \neq \emptyset$ also imply $I = R$. Hence $I \subseteq 2R$ or $I \subseteq 3R$.

(b) From (a) follows that $rad(R) = 2R \cap 3R = 6R$ which is surely not a nil ideal (see also 2.19) and $R/rad(R) = R/6R \cong (R/2R) \oplus (R/3R)$ a direct sum (see 2.24) of simple rings (fields), i.e. a semisimple ring. Finally, if we consider in $R/6R$ the idempotent $3 + 6R$ there is no element x idempotent in R (neither in $\mathbb{Q}$ except for 0 and 1) such that $x + 6R = 3 + 6R$: the idempotents do not lift modulo $rad(R)$.

Chapter 13

Prime Ideals, Local Rings

Ex. 13.1 We have already used $N(xy) = N(x).N(y)$ in the previous chapter, for the "norm" $N(a + bi) = a^2 + b^2$ in $\mathbb{Z}[i]$. The units in $\mathbb{Z}[i]$ being $\{\pm 1, \pm i\}$ i.e. the elements $x \in \mathbb{Z}[i]$ with $N(x) = 1$ we infer that x is prime in $\mathbb{Z}[i]$ iff $N(x)$ is prime in $\mathbb{Z}$ (indeed, the non-associated decompositions correspond to each other). Hence $N(1 + i) = 2$ implies that the ideal $(1 + i)$ is prime.

Further, we can prove that a prime number $p \in \mathbb{N}^*$ is also prime in $\mathbb{Z}[i]$ iff the equation $a^2 + b^2 = p$ has no solutions in $\mathbb{Z}$. Indeed, if $a^2 + b^2 = p$ has integer solutions then $p = (a + bi)(a - bi)$ is a non-trivial decomposition because $N(a + bi) = N(a - bi) = p \neq 1$. Conversely, if $p = (a + bi)(c + di)$ is a non-trivial decomposition, from $N(a+bi).N(c+di) = p^2$ we must have $N(a+bi) = N(c+di) = p$. The product $(a + bi)(c + di)$ being real and $a + bi, c + di$ having the same module, $c + di = a - bi$. Hence $p = a^2 + b^2 = (a + bi)(a - bi)$.

Finally, $a^2 + b^2 = 3$ clearly has no integer solutions but $a^2 + b^2 = 2$ has $a = b = 1$. Hence (3) is prime but (2) is not.

Ex. 13.2 $\mathbb{Z}$ being a commutative ring, an ideal $n\mathbb{Z}$ is prime iff $\mathbb{Z}/n\mathbb{Z}$ is an integral domain, i.e. iff $\mathbb{Z}_n$ is an integral domain. But this happens iff n is a prime number. Hence $2\mathbb{Z}, 3\mathbb{Z}, 5\mathbb{Z}, .., p\mathbb{Z}, ..$ (p prime) are the prime ideals in $\mathbb{Z}$.

Ex. 13.3 Obviously each simple ring is prime so that, being simple (see 2.20), $\mathcal{M}_n(K)$ is prime. Now, $i^{-1}(0) = 0$ (i being injective) but the subring $U_{21}(K)$ is not prime: indeed $I = \begin{pmatrix} 0 & K \\ 0 & K \end{pmatrix}$ and $J = \begin{pmatrix} 0 & K \\ 0 & 0 \end{pmatrix}$ are nonzero left ideals in $U_{21}(K)$ (I is also a left ideal in $M_2(K)$ but J is not) but $I \circ J = 0$.

Ex. 13.4 Let A, B be ideals in R such that $A \not\subseteq I$ and $B \not\subseteq I$. By maximality of I, $A + I$ and $B + I$ must meet S, that is, there are elements $s_1 \in (A + I) \cap S$ and $s_2 \in (B + I) \cap S$. S being multiplicative $s_1 s_2 \in S \cap (A + I)(B + I)$ and so $s_1 s_2 \in AB + I$. Hence $s_1 s_2 \notin I$ implies $AB \not\subseteq I$.

Ex. 13.5 In order to prove $\sqrt{I} = \cap \{P \text{ prime ideal} \mid I \subseteq P\}$ we shall denote the right member by A and verify both inclusions. If $I \subseteq P$ and $a \in \sqrt{I}$ then for a suitable $m \in \mathbb{N}^*$, $a^m \in I$ or $a^m \in P$ and hence $a \in P$ (P being prime). So $\sqrt{I} \subseteq A$. Conversely, let $a \notin \sqrt{I}$. The set $X = \{a, a^2, .., a^n, ..\}$ is then disjoint from I and is clearly a multiplicative set ($0 \notin X$). As in the previous exercise (one has to apply in a similar way Zorn's lemma), there is a prime ideal P maximal with the properties $P \cap X = \emptyset$ and $I \subseteq P$. Hence $a \notin P$ and $a \notin A$.

Ex. 13.6 (a) The subjacent additive group of K_4 being of type Klein, we have $x + x = 0$ for each $x \in K_4$ so that $char(K_4) = 2$.

(b) Clearly $1 + 1 = 0$ so that the prime subfield $P(K_4) = \{0, 1\} \cong \mathbb{Z}_2$. If we denote by $0, 1, a, b$ the elemnts of K_4 then $a^2 = a + 1$ and $b^2 = b + 1$ follow by simple verifications.

(c) Being a complex cubic root of the unity, $w \notin \mathbb{R}$, we have $w^2 + w + 1 \doteq 0$ so that in $\mathbb{Z}[w]$ multiplication is defined as follows: $(a + bw)(c + dw) = ac - bd + (ad + bc - bd)w$. One can easily check directly that $x, y \notin (2)$ implies $x \cdot y \notin (2)$ and hence (2) is prime.

(d) If $x = a + bw$ then $x \in (2) = 2 \cdot \mathbb{Z}[w] \Leftrightarrow a, b \in 2\mathbb{Z}$. The corresponding congruence relation being defined as $x \equiv y(mod(2)) \Leftrightarrow x - y \in (2)$ the quotient ring will have the following four classes: $2\mathbb{Z}[w]$, $2\mathbb{Z} + (2\mathbb{Z} + 1)w$, $(2\mathbb{Z} + 1) + 2\mathbb{Z}w$, $(2\mathbb{Z} + 1) + (2\mathbb{Z} + 1)w$. The ideal (2) being prime the factor ring $\mathbb{Z}[w]/(2)$ is an integral domain. Choosing as representatives $0, 1, w, 1 + w$ the function $\begin{pmatrix} 0 & 1 & a & b \\ \bar{0} & \bar{1} & \bar{w} & \overline{1+w} \end{pmatrix}$ is the required isomorphism.

Ex. 13.7 According to 2.5 all the ideals in $4\mathbb{Z}$ have the form $4n\mathbb{Z}$ so that clearly, $(8) = 8\mathbb{Z}$ is a maximal ideal. But $4\mathbb{Z}/8\mathbb{Z} = \{8\mathbb{Z}, 4 + 8\mathbb{Z}\}$ has obviously no identity (it is a zero square ring), and hence it is not a field.

Ex. 13.8 Consider the map $f : M_2(\mathbb{Z}) \to M_2(\mathbb{Z}/p\mathbb{Z})$ defined by $f\left(\left(\begin{array}{cc} a & b \\ c & d \end{array}\right)\right) = \left(\begin{array}{cc} [a]_p & [b]_p \\ [c]_p & [d]_p \end{array}\right)$. One verifies that this is a surjective ring homomorphism and $\ker f$ is precisely A. But $\mathbb{Z}/p\mathbb{Z}$ being a field, $M_2(\mathbb{Z}/p\mathbb{Z})$ has no proper ideals (see 2.20) and so, using an isomorphism theorem $M_2(\mathbb{Z}/p\mathbb{Z}) \cong M_2(\mathbb{Z})/H$ and hence H is a maximal ideal in $M_2(\mathbb{Z})$. Surely, together with $M_2(\mathbb{Z}/p\mathbb{Z})$, $M_2(\mathbb{Z})/H$ is no field.

Ex. 13.9 If M is maximal and $x \notin M$ then $M < M + Rx$ and hence $M + Rx = R$. From $1 \in M$ there is an element $r \in R$ such that $1 = m + rx$ for a suitable $m \in M$, i.e. $1 - rx \in M$. Conversely, if $M < N \triangleleft R$ and $x \in N \backslash M$ then there is an element $r \in R$ such that $1 - rx \in M$ and hence $1 - rx \in N$. But then $1 \in N$, $N = R$ and M is maximal.

Ex. 13.10 Consider the abelian group $(\mathbb{Z}_p, +)$ together with the zero multiplication $\bar{a}.\bar{b} = \bar{0}, \forall \bar{a}, \bar{b} \in \mathbb{Z}_p$. Obviously, this is a simple commutative ring (the ideals are precisely the subgroups of $\mathbb{Z}_p$) and $(\bar{0})$ is a maximal but not prime ideal ($\bar{1}.\bar{1} = \bar{0}$ does not imply $\bar{1} = \bar{0}$).

Ex. 13.11 If $a \notin \mathcal{R}(R)$ there is a prime ideal P such that $a \notin P$. The ideal P being prime $a^2 \notin P$ and similarly $a^3, .., a^n, .. \notin P$. Hence a is not nilpotent, $a \notin \mathcal{N}(R)$ and $\mathcal{N}(R) \subseteq \mathcal{R}(R)$. Conversely, let $a \notin \mathcal{N}(R)$ and $X = \{a, a^2, .., a^n, ..\}$. Then $0 \notin X$ and let P be an ideal maximal with the property $P \cap X = \emptyset$. If we prove that P is prime (see also 13.4) then $a \notin P$ implies $a \notin \mathcal{R}(R)$ and hence $\mathcal{R}(R) \subseteq \mathcal{N}(R)$.

Let I, J be ideals in R such that $I \not\subset P$ and $J \not\subset P$. By maximality of P, $I + P$ and $J + P$ must meet X and hence there are $n, m \in \mathbb{N}^*$ such that $a^n \in I + P, a^m \in J + P$. So $a^{\max(n,m)} \in (I + P)(J + P) \subseteq IJ + P$ and $a^{\max(n,m)} \notin P$ implies $IJ \not\subset P$.

Ex. 13.12 If I, J are ideals in a ring R and $I \subseteq J$ then, using a well-known isomorphism theorem $R/J \cong (R/I)/(J/I)$, we deduce that J is a prime ideal in R iff J/I is a prime ideal in R/I. Considering $I = \mathcal{R}(R)$ we infer $\mathcal{R}(R/\mathcal{R}(R))) = (0)$.

Ex. 13.13 If T denotes the intersection of all the maximal ideals of $R[X]$ suppose $0 \neq f \in T$. Let M be a maximal ideal which includes the principal ideal $(1 + X \cdot f)$. Such a maximal ideal exists because $f \neq 0$ implies $1 + X \cdot f$ is not a unit in $R[X]$ (see also 14.15). But together with $f \in M$ we must have $1 \in M$, a contradiction.

Ex. 13.14 Let M be the ideal of all the polynomials with zero constant term and let A be an ideal of $\mathbb{C}[X, Y]$ which contains strictly M. The ideal A must contain a polynomial with a non-zero constant term $a \neq 0$, say $f = a + \sum_{i,j} a_{ij} X^i Y^j$. But clearly the polynomial $f - a = \sum_{i,j} a_{ij} X^i Y^j \in M \subset A$ and hence $0 \neq a \in A$. In $\mathbb{C}[X, Y]$ the units are the non-zero complex numbers (e.g. a) so that $A = \mathbb{C}[X, Y]$ (an ideal which contains a unit).

Ex. 13.15 $\mathbb{Q}[X, Y]$ being a commutative ring an ideal P is prime iff $f \cdot g \in P$ implies $f \in P$ or $g \in P$. Moreover, being with identity, each maximal ideal is prime.

$X \cdot X \in (X^2)$ but $X \notin (X^2)$ so that (X^2) is neither prime nor maximal.

$(X - 2, Y - 3)$ is maximal and so also prime (one makes a similar judgement as in the previous exercise).

$(Y - 3)$ is prime (each polynomial in this ideal is a polynomial in $Y - 3$ with zero constant term; if a product has this form at least one factor has also this form). It is strictly included in the previous ideal, and hence it is not a maximal ideal.

$(X^2 + 1)$ is prime because $X^2 + 1$ is irreducible over $\mathbb{Q}$ but it is not a maximal ideal (being strictly included in $(X^2 + 1, Y - 3)$). We have $X^2 - 1 = (X - 1)(X + 1)$ so this ideal is neither prime nor maximal.

As above $(X^2 + 1, Y - 3)$ is a prime ideal. It is not a maximal ideal being strictly included in $(X^2 + 1, Y - 3, X^3) \neq \mathbb{Q}[X, Y]$ (the polynomials of first degree in X do not belong to this ideal).

Ex. 13.16 According to 14.19 we can use the isomorphism $\mathbb{Z}[X]/(n, X) \cong \mathbb{Z}_n$. Hence our statement is equivalent to $\mathbb{Z}_n$ is an integral domain iff n is prime, a well-known result.

Ex. 13.17 Using a well-known result of Jacobson, such a ring is commutative. Now, if P is a prime ideal then the quotient ring R/P is an integral domain. In order to prove that P is also maximal we have to verify that R/P actually is a field. Indeed, if $x \in R \backslash P$ we have $(x + P)\left(x^{n(x)-1} + P\right) = x^{n(x)} + P = x + P = (x + P)(1 + P)$ and hence, R/P having no zero divisors, $x^{n(x)-1} + P = 1 + P$. If $n(x) = 2$ then $x + P = 1 + P$ is obviously a unit and if $n(x) > 2$ then $x^{n(x)-2}$ is inverse for $x + P$. The application is obvious: in a Boole ring $\forall x \in R : n(x) = 2$.

Ex. 13.18 (a) $V(A) \subseteq V(E)$ follows from $E \subseteq A$. Conversely, $P \in V(E)$ implies $E \subseteq P$ and $(E) \subseteq P$ or $A \subseteq P$. Hence $P \in V(A)$.
   (b) Obvious.

   (c) Indeed, $P \in V\left(\bigcup_{i \in I} E_i\right) \Leftrightarrow \bigcup_{i \in I} E_i \subseteq P \Leftrightarrow \forall i \in I : E_i \subseteq P \Leftrightarrow$
$\forall i \in I : P \in V(E_i) \Leftrightarrow P \in \bigcap_{i \in I} V(E_i)$.

   (d) From $A \circ B \subseteq A \cap B$ we deduce $V(A \cap B) \subseteq V(A \circ B)$. Further, $P \in V(A \circ B) \Leftrightarrow A \circ B \subseteq P$ and P being a prime ideal, $\Leftrightarrow A \subseteq P$ or $B \subseteq P \Leftrightarrow P \in V(A) \cup V(B)$ that is $V(A \circ B) = V(A) \cup V(B)$. Finally, $P \in V(A) \cup V(B) \Leftrightarrow A \subseteq P$ or $B \subseteq P \Rightarrow A \cap B \subseteq P \Leftrightarrow P \in V(A \cap B)$ that is $V(A) \cup V(B) \subseteq V(A \cap B)$ and hence all the inclusions are equalities.

   (e) Indeed, the family $\mathcal{V}$ of subsets $V(E)$ for each $E \in \mathcal{P}(R)$, satisfies the axioms of a topological space defined in terms of the closed subsets: $\mathcal{V}$ is closed under arbitrary intersections (from (c), $\mathcal{V}$ is closed under finite reunions (from (a) and (d)) and $\emptyset$ and $Spec(R)$, belong to $\mathcal{V}$ (from (b)).

Ex. 13.19 From the previous chapter (see 13.2) we already know that $Spec(\mathbb{Z}) = \{p\mathbb{Z} | p$ is a prime number$\}$ and $Spec(\mathbb{R}) = \{0\}$ because every division ring has no proper ideals. Moreover, the closed sets in $Spec(\mathbb{Z})$ are the sets $V(E) = \{p\mathbb{Z} | E \subseteq p\mathbb{Z}\}$ for all $E \in \mathcal{P}(\mathbb{Z})$. But $V(E) = V((E))$ and, all the ideals in $\mathbb{Z}$ having the form $n\mathbb{Z}$, the closed sets are only the finite sets $\{p_k \mathbb{Z} | 1 \le k \le m, p_k$ prime number$\}$. Indeed, if $n = p_1^{r_1} p_2^{r_2} .. p_m^{r_m}$ then the only prime ideals which contain $n\mathbb{Z}$ are $p_1 \mathbb{Z}, p_2 \mathbb{Z}, .., p_m \mathbb{Z}$. On $Spec(\mathbb{R})$ we obviously have the indiscrete topology.

Ex. 13.20 From an above exercise we deduce immediately that each open set $Spec(R) \backslash V(E) = \bigcup_{r \in E} U_r$, so the U_r's form a base for the open sets in the Zariski toplogy.

(a) Indeed, denoting $S = Spec(R)$ we have $U_r \cap U_s = (S \backslash V(r)) \cap (S \backslash V(s)) = S \backslash (V(r) \cup V(s)) = \{P \in S | \{r, s\} \not\subseteq P\} = \{P \in S | r \notin P$ and $s \notin P\} = \{P \in S | rs \notin P\} = S \backslash V(rs) = U_{rs}$.

(b) $U_r = \emptyset$ iff $V(r) = Spec(R)$ iff $\forall P \in Spec(R) : r \in P$ iff $r \in \mathcal{R}(R)$ the prime radical of R. In a commutative ring (see 13.11) $\mathcal{R}(R)$ is precisely the set of all the nilpotent elements of R.

(c) $U_r = Spec(R)$ iff $V(r) = \emptyset$. But $V(r) = V((r))$ and $(r) = R$ holds iff r is a unit so that $V(r) = \emptyset$ iff r is a unit (one uses also the fact that each proper ideal is included in a maximal (and hence prime) ideal of R).

(d) Surely $U_r = U_s$ iff $V(r) = V(s)$. According to 13.5 $\sqrt{(r)}$ is the intersection of all the prime ideals which contain r and hence $V(r) = V(s) \Leftrightarrow \sqrt{(r)} = \sqrt{(s)}$ (one checks that for two ideals A, B of R the inclusion $V(A) \subseteq V(B)$ is equivalent with $B \subseteq \sqrt{A}$).

(e) Obviously $Spec(R) = U_1$(i.e. $r = 1$) so it suffices to deal with the subsets U_r. In order to check that U_r is quasi compact, $\{U_r | r \in R\}$ being a base for the open sets in the Zariski topology, it suffices to show that if $\{r_i | i \in I\} \in \mathcal{P}(R)$ and $U_r \subseteq \bigcup_{i \in I} U_{r_i}$ there is a finite subset $F \subseteq I$ such that $U_r \subseteq \bigcup_{i \in F} U_{r_i}$. Indeed, $U_r \subseteq \bigcup_{i \in I} U_{r_i} \Leftrightarrow V(r) \supseteq \bigcap_{i \in I} V(r_i) \Leftrightarrow V(r) \supseteq V\left(\bigcup_{i \in I} \{r_i\}\right) \Leftrightarrow (r) \subseteq \sqrt{(\{r_i | i \in I\})}$. But this last condition implies the existence of a $n \in \mathbb{N}^*$ such that $r \in (\{r_i | i \in I\})$. Hence there is a finite subset $F \subseteq I$ such that $r \in (\{r_i | i \in F\})$ or $U_r \subseteq \bigcup_{i \in F} U_{r_i}$.

Ex. 13.21 (a) Indeed, $(f^*)^{-1}(U_r) = \{P' \in Y | f^*(P') \in U_r\} = \{P' \in Y | f^{-1}(P') \in U_r\} = \{P' \in Y | f^{-1}(P') \notin V(r)\}$

$= \{P' \in Y | r \notin f^{-1}(P')\} = \{P' \in Y | f(r) \notin P'\} = U_{f(r)}$. The subsets U_r being a base for the open sets we deduce that f^* is continuous.

(b) A similar computation shows that the required equality is equivalent with $\{P' \in Y | A \subseteq f^{-1}(P')\} = \{P' \in Y | f(A).R' \subseteq P'\}$. Using a previous property $V(f(A)) = V(f(A).R')$ and hence all reduces to $A \subseteq f^{-1}(P') \Leftrightarrow f(A) \subseteq P'$, which is clear.

(c) Since $f^*(V(A')) = f^{-1}(V(A')) = \{f^{-1}(P')) \in X | A' \subseteq P'\}$ we only have to show that $\overline{f^*(V(A'))} = \cap\{V(E) | f^{-1}(V(A')) \subseteq V(E)\} = V(f^{-1}(A'))$. Hence we must verify:

1^0 $f^{-1}(V(A')) \subseteq V(E) \Rightarrow V(f^{-1}(A')) \subseteq V(E)$, and

2^0 $f^{-1}(V(A')) \subseteq V(f^{-1}(A'))$. In order to verify the first implication one needs to show that $[\forall P' \in Y, A' \subseteq P' \Rightarrow E \subseteq f^{-1}(P')] \Rightarrow [\forall P \in X, f^{-1}(A') \subseteq P \Rightarrow E \subseteq P]$. If A' is a prime ideal this is clear. If A' is not prime let us consider $\sqrt{A'}$ which is (in a commutative ring, see 13.5) the intersection of all the prime ideals including A'. We surely have $A' \subseteq \sqrt{A'}$ and hence by hypothesis $E \subseteq f^{-1}\left(\sqrt{A'}\right)$. But $f^{-1}\left(\sqrt{A'}\right) = \sqrt{f^{-1}(A')}$ is easily checked and so from $f^{-1}(A') \subseteq P \Rightarrow \sqrt{f^{-1}(A')} \subseteq P$ we deduce $E \subseteq f^{-1}\left(\sqrt{A'}\right) = \sqrt{f^{-1}(A')} \subseteq P$. Finally, 2^0 follows immediately from $A' \subseteq P' \Rightarrow f^{-1}(A') \subseteq f^{-1}(P')$.

(d) Clearly, if f is surjective and $\ker(f) \subseteq P$ then $f(P) \in Y$. Hence we can define a map $g^* : V(\ker(f)) \to Y$ by $g^*(P) = f(P)$. From (c) we remark that if $A' = (0)$ then $\overline{f^*(Y)} = \overline{f^*(V(0))} = V(f^{-1}(0)) = V(\ker(f))$ and so $f^*(Y) \subseteq \overline{f^*(Y)} = V(\ker(f))$. Hence $(g^* \circ f^*)(P') = g^*(f^{-1}(P')) = (f \circ f^{-1})(P') = P'$, because f is surjective, and so $g^* f^* = 1_Y$ or f^* is injective. All the ideals of R which include $\ker(f)$ having the form $f^{-1}(A')$, f^* is also surjective (we compute the values $f^*(V(A')) = V(f^{-1}(A')))$.

(e) Follows from (c) because $\overline{f^*(Y)} = \overline{f^*(V(0))} = V(f^{-1}(0)) = V(0) = X$ ($f^{-1}(0) = (0)$ follows from the injectivity of f).

Ex. 13.22 (a) In a Boole ring each element is idempotent and $\forall r \in R, r^2 = r \Rightarrow r(1-r) = 0$. Using a previous property we deduce $U_r \cap U_{1-r} = U_{r(1-r)} = U_0 = \emptyset$. Further, $U_r \cup U_{1-r} = Spec(R) \backslash V(\{r, 1-r\}) = Spec(R) \backslash V((r, 1-r)) = Spec(R) \backslash V(1) = Spec(R0$ and so $U_r = Spec(R) \backslash U_{1-r}$ is also closed.

(b) We use a previous result (see 2.10): in a Boole ring each finitely generated ideal is principal.

(c) Let $A \subseteq Spec(R)$ an open and closed set. As open set $A = \bigcup_{i \in I} U_{r_i}$ a union of elements in the base of open sets. A being closed in a quasicompact (see a previous exercise) topological space, is quasicompact itself. Then the above cover of A has a finite subcover $A = \bigcup_{i \in F} U_i$ for a finite subset $F \subseteq I$. Now our result follows from (b).

(d) We remark that the closure of a point $P \in Spec(R)$ is simply $\overline{\{P\}} = V(P)$. Hence $\{P\}$ is closed iff P is maximal. In a Boole ring, the prime ideal being all maximal (see 13.17), "the points" are closed in $Spec(R)$, so that $Spec(R)$ is Hausdorff. Using (e) from a previous exercise we obtain the information required: $Spec(R)$ is a Hausdorff compact topological space.

Ex. 13.23 First observe that each ideal from R which contains a unit is equal to R. Then if N is the set of all the non-units of R, all the proper ideals in R are included in N.

Now, let M be the unique maximal ideal of R. We show that $M = N$. From above $M \subseteq N$ is clear. If $r \in N$ then the ideal Rr is included in N and hence is proper. But each proper ideal must be included in a maximal ideal, and this is only M. From $Rr \subseteq M$ we infer $r \in M$ (R has identity) and hence $N \subseteq M$. Conversely, if N is an ideal it is clearly also the unique maximal ideal of R.

Ex. 13.24 From 2.27 we know that $\mathbb{Q}^{(p)}$ has a descending chain of ideals of the form $\mathbb{Q}^{(p)} \supset p\mathbb{Q}^{(p)} \supset p^2\mathbb{Q}^{(p)} \supset ..$ Hence $\mathbb{Q}^{(p)}$ has a unique maximal ideal (the set of non-units) and consequently (see also 13.23), is a local ring.

Ex. 13.25 Let $a \in R \backslash M$. The ideal M being maximal, the quotient ring R/M is a field and there is an element $b \in R$ such that $(a + M)(b + M)' = 1 + M$ and hence $c = ab - 1 \in M$. Then $ab = c + 1$ is a unit (see 13.9) and so is a. But then M is the set of all the non-units (the converse inclusion is always valid, 13.9 again) of R and R is a local ring.

Ex. 13.26 Let $a \in M$. We have $(a+1)(1-a+a^2-..+(-1)^{n-1}a^{n-1}) = 1 + (-1)^n a^n$ and hence the coset $(a+1) + M^n$ is a unit in R/M^n. A well-known theorem of isomorphism $(R/M^n)/(M/M^n) \cong R/M$ proves that M/M^n is also a maximal ideal in R/M^n. But then , using the previous exercise, R/M^n is a local ring.

Chapter 14

Polynomial Rings

Ex. 14.1 The polynomial $f = X^2 - \bar{1}$ has 4 zeros in $\mathbb{Z}_{15}$: $\bar{1}, \bar{4}, \overline{11}, \overline{14}$.

Ex. 14.2 The polynomial $f = X^2 + 1$ has an infinity of zeros in $\mathbf{H}$ (see 7.12). Indeed, all the quaternions $\alpha = b\mathbf{i} + c\mathbf{j} + d\mathbf{k}, b^2 + c^2 + d^2 = 1$ are zeros for f.

Ex. 14.3 A standard computation goes like this:

$x^2 + px + q = \left(x + \frac{p}{2}\right)^2 + q^2 - \left(\frac{p}{2}\right)^2 = 0$ and hence $x = -\frac{p}{2} + \sqrt{\left(\frac{p}{2}\right)^2 - q}$.

In order that $\frac{p}{2}$ (the solution of $y + y = p$) to exist, the (commutative) field must not have the characteristic 2. Moreover, each element in the field has to be a square (and $\sqrt{\left(\frac{p}{2}\right)^2 - q}$ denotes here any solution of the equation $z^2 = \left(\frac{p}{2}\right)^2 - q$).

Ex. 14.4 One has to show that $h_1 + (f) = h_2 + (f) \Rightarrow h_1 + (g) = h_2 + (g)$ in order to verify that F is well-defined. But g divides f and hence $f | h_1 - h_2$ implies $g | h_1 - h_2$. F is clearly a surjective ring homomorphism so that $(\mathbb{Q}[X]/(f))/\ker(F) \cong \mathbb{Q}[X]/(g)$. But g is an irreducible polynomial in $\mathbb{Q}$ and hence $\mathbb{Q}[X]/(g)$ is a field and $\ker(F)$ is a maximal ideal in $\mathbb{Q}[X]/(f)$.

Ex. 14.5 Consider the map $F : R[X] \to R, F(a_0 + a_1 X + .. + a_n X^n) = a_0$ which surely is a surjective ring homomorphism such that $\ker(F) = (X)$. Then $R \cong R[X]/(X)$ and $R[X]/(I)$ being a field, (I) is a maximal ideal.

Ex. 14.6 Suppose that $a.D[X] + X.D[X] = (f)$ for a $f \in D[X]$. Then from $a = fg, g \in D[X]$ follows that $a \in D$ and from $X = fg_1, g_1 \in D[X]$ we deduce that f and then a (associated with f) is a unit (indeed if $1 = ua + vX$ then $0 = (u - u_0)a + vX$ and $1 = u_0a$ where u_0 is the constant term of $u \in D[X]$), a contradiction.

Ex. 14.7 One uses the decompositions: $X^3 + X^2 + X + \bar{1} = (X^2 + \bar{1})(X + \bar{1})$ and $X^2 + \bar{3}X + \bar{2} = (X + \bar{2})(X + \bar{1})$. Yes, for $\mathbb{Z}_5[X]$.

Ex. 14.8 $X^2 + \bar{1}$ is irreducible in $\mathbb{Z}_3[X]$ but $X^2 + \bar{1} = (X + \bar{2})(X + \bar{3})$ holds in $\mathbb{Z}_5[X]$. The second polynomial is reducible in both $\mathbb{Z}_3[X]$ and $\mathbb{Z}_5[X]$: indeed $X^3 + X + \bar{2} = (X + \bar{1})(X^2 - X + \bar{2})$.

Ex. 14.9 $\mathbb{Z}_5$ being a field, $(\mathbb{Z}_5^*, \cdot)$ is a group with 4 elements. Using the well-known Lagrange theorem we obtain $t^4 = \bar{1}$ for each $\bar{0} \neq t \in \mathbb{Z}_5$. Hence $f(t) = x^4 + at + \bar{1} = at + \bar{2}$ and obviously for each $a \neq \bar{0}$ the polynomial has roots in $\mathbb{Z}_5 (a = \bar{1}, t = \bar{3}; a = \bar{2}, t = \bar{4}$ and conversely). In the case $a = \bar{0}$ one finds $f = X^4 + \bar{1} = (X^2 + \bar{2})(X^2 + \bar{3})$ by usual coefficient identification. So f is always reducible.

Ex. 14.10 One simply verifies $f(\bar{0}) = g(\bar{0}) = \bar{0}, f(\bar{1}) = g(\bar{1}) = \bar{0}$ and $f(\bar{2}) = g(\bar{2}) = \bar{0}$.

Ex. 14.11 Consider the map $f : R[X] \to R$ defined by $f(a_0 + a_1X + .. + a_nX^n) = a_0$ which obviously is a surjective ring homomorphism. Then for an ideal I in R we have $I_* = f^{-1}(I)$ which is an ideal in $R[X]$. Moreover, $R[X]/I_* \cong R/I$ (just take the composition $R[X] \xrightarrow{f} R \xrightarrow{p_I} R/\dot{I}$ and apply a theorem of isomorphism, $\ker(p_I \circ f) = I_*$) and so the assertions concerning I and I_* prime (resp. maximal) follow from the well-known characterizations of these using the quotient ring (integral domain resp. field) of a commutative ring with identity.

Ex. 14.12 If $n \neq 0$ and $(n, X) = (f), f \in \mathbb{Z}[X]$ then $\exists p, q \in \mathbb{Z}[X]$ such that $n = f \cdot p, X = f \cdot q$. $\mathbb{Z}$ being an integrity domain $0 = \deg(f) + \deg(p)$ and hence $\deg(f) = 0$ or $f = a \in \mathbb{Z}$. Then from $1 = \deg(f) + \deg(q)$ we deduce $\deg(q) = 1$ or $q = bX + c$, where $b, c \in \mathbb{Z}$. From $X = a \cdot (bX + c)$ we obtain $c = 0$ and $a \in U(\mathbb{Z})$ so that $a \in \{-1, 1\}$ and $n \in \{-1, 1\}$. Conversely, for $n = 0$ the ideal $(n, X) = (X)$ is clearly principal. For $n \in \{-1, 1\}$ one has $(n, X) = \mathbb{Z}[X]$ also a principal ideal.

Ex. 14.13 The first verification (ideal) is easy. Moreover, this ideal is in fact $I = (2, X)$. If $I = (f)$ then f would be a common divisor of 2 and X and hence $f \in \{-1, 1\}$. But this implies $I = \mathbb{Z}[X]$, impossible.

Remark. Unless R is a field, $R[X]$ is not necessarily a principal ring.

Ex. 14.14 If $T_1 = \{f = a_0 + a_1 X + .. + a_n X^n \in R[X] | a_1 = 0\}$ then one easily checks the implication $f, g \in T_1 \Rightarrow f + g, f \cdot g \in T_1$. Further, if $T_2 = \{f \in R[X] | a_2 = 0\}$, this is no subring because for instance $X, X + X^3 \in T_2$ but $X \cdot (X = X^3) = X^2 + X^4 \notin T_2$.

Ex. 14.15 Let $f \in R[X], f = a_0 + a_1 X + .. + a_n X^n$. In what follows we prove that:

(i) f is a unit in $R[X]$ iff a_0 is a unit in R and $a_1, a_2, .., a_n$ are nilpotent elements in R;

(ii) f is nilpotent in $R[X]$ iff all the coefficients are nilpotent.

In order to simplify the notation $U(R)$ and $U(R[X])$ denote the corresponding sets of units.

(i) First observe that in a commutative ring a finite sum of nilpotent elements is also nilpotent (see 2.15). Further, if $a_0 \in U(R)$ and $a_1, .., a_n$ are nilpotent then $a_1 X + a_2 X^2 + .. + a_n X^n$ is clearly nilpotent in $R[X]$ and then $f \in U(R[X])$ (see 1.25). Conversely, if $f \in U(R[X]), f \cdot g = 1$ and $g = b_0 + b_1 X + .. + b_m X^m$ then $a_0 b_0 = 1, a_0 b_1 + a_1 b_0 = 0, .., a_{n-1} b_m + a_n b_{m-1} = 0, a_n b_m = 0$ hold. Multiplying the last equality by a_n and using the previous one gets $a_n^2 \cdot b_{m-1} = 0$. Multiplying the next equality (in reversed order) by a_n^2 we obtain $a_n^3 b_{m-2} = 0$ and so on. Finally, $a_n^{m+1} b_0 = 0$ implies $a_n^{m+1} = 0$ (multiplying by a_0) and hence a_n is nilpotent in R. Using again 1.25 $f - a_n X^n = a_0 + a_1 X + .. + a_{n-1} X^{n-1} \in U(R[X])$ and we deduce that a_{n-1} is nilpotent (and $a_{n-2}, .., a_2, a_1$) in an analogous way. Surely, $a_0 \in U(R)$ follows from $a_0 b_0 = 1$.

(ii) Using again 1.25, if $f \in R[X]$ is nilpotent then $1 - f \in U(R[X])$ and hence by (i) $(1 - a_0 \in U(R)$ and$)$ $-a_1, -a_2, .., -a_n$ are nilpotent. But then (using again f nilpotent for a_0) $a_0, a_1, .., a_n$ are nilpotent. The converse is obvious.

Ex. 14.16 (a)$\Rightarrow$(b) Let $f = a_0 + a_1 X + .. + a_n X^n$ and $g = b_0 + b_1 X + .. + b_m X^m$ be polynomials in $R[X]$ (with $a_n \neq 0 \neq b_m$) and $f \cdot g = 1$. Because $a_0 b_0 = 1$ one can suppose (modifying the coefficients of f and

g) that $a_0 = b_0 = 1$. A simple induction shows that $\forall k \in \mathbb{N}, k \le m$ the equality $a_n^{k+1} b_{m-k} = 0$ (indeed $a_n b_m = 0$ holds by the definition of polynomial multiplication. If $a_n b_m = a_n^2 b_{m-1} = .. = a_n^k b_{m-k+1} = 0$ the required relation follows from $a_{n-k} b_m + .. + a_{n-k+h} b_{m-h} + .. + a_n b_{m-k} = 0$ multiplying by a_n^k). Now, from $a_n^{m+1} b_0 = 0$ (and $b_0 = 1$) we have $a_n^{m+1} = 0$ and hence $a_n = 0$, a contradiction.

(b)$\Rightarrow$(a) If $a \in R$ is a nonzero nilpotent element and $a^n = 0, a^{n-1} \ne 0, n \in \mathbb{N}$ then $R[X]$ has units of degree at least 1. Indeed, $(1 + a^{n-1} X)(1 - a^{n-1} X) = 1 - a^{2n-2} X^2 = 1$.

Ex. 14.17 $\mathbb{C}[X]$ being a commutative ring, $\mathbb{C}[X]/(X^2+1)$ is an integral domain iff $(X^2 + 1)$ is a prime ideal. But $X^2 + 1 = (X + i)(X - i) \in (X^2 + 1)$ but neither $X + i$ nor $X - i$ does not belong to the principal ideal $(X^2 + 1)$.

Another solution: denoting by $U = (X^2 + 1)$ the equality $[(X + i) + U] \cdot [(X - i) + U] = U$ points out zero divisors in $\mathbb{C}[X]/U$.

Ex. 14.18 Let $E_i : \mathbb{R}[X] \to \mathbb{C}$ be the surjective ring homomorphism defined by $E_i(f) = a_0 + a_1 i + a_2 i^2 + .. + a_n i^n, \forall f = a_0 + a_1 X + .. + a_n X^n \in \mathbb{R}[X]$ (such a ring homomorphism exists in more generally conditions: if R' is a ring with identity and R is a subring of R' which contains the identity, then for each element $c \in Z_{R'}(R)$ the centralizer, there is a unique unital ring homomorphism $E_c : R[X] \to R'$ such that $E_c(a) = a$ for each $a \in R$ and $E_c(X) = c$; one takes $\mathbb{R}$ subring in $\mathbb{C}$ and $i \in Z_{\mathbb{C}}(\mathbb{R})$). Obviously, $\ker(E_i) = (X^2 + 1)$ (the ideal generated by $X^2 + 1$) and by an isomorphism theorem $\mathbb{R}[X]/\ker(E_i) \cong E_i(\mathbb{R}[X])$ that is, the required isomorphism.

Ex. 14.19 Let us consider the following ring homomorphisms:
$$\mathbb{Z}[X] \overset{q}{\to} \mathbb{Z} \overset{p}{\to} \mathbb{Z}_n \text{ where } q(a_0 + a_1 X + .. + a_n X^n) = a_0 \text{ and } p(x) = \bar{x}, \forall x \in \mathbb{Z}.$$ The following equalities hold: $\ker(q) = (X), \ker(p) = (n)$ and $\ker(p \circ q) = (n, X) = (n) + (X)$ (surely $(n) \subseteq \mathbb{Z}[X]$). The required isomorphism is obtained from an isomorphism theorem (q and p are surjective): $\mathbb{Z}[X]/\ker(p \circ q) \cong (p \circ q)(\mathbb{Z}[X])$.

Chapter 15

Rings of Quotients

Ex. 15.1 $\mathbb{Z}$ being an integral domain, for each multiplicative system $S \subset \mathbb{Z}$, the ring of quotients is identified with a subring with identity of $\mathbb{Q}$, the rational numbers of the form $\frac{m}{s}$ with $m \in \mathbb{Z}$ and $s \in S$. Now, if A is a subring of $\mathbb{Q}$ and $1 \in A$ let $S_A = \left\{ n \in \mathbb{Z} | \frac{1}{n} \in A \right\}$. Obviously, S_A is a multiplicative system and $\mathbb{Z}_{S_A} \subseteq \mathbb{Q}$. Conversely, if $\frac{m}{n} \in A$ with $(m; n) = 1$ let $u, v \in \mathbb{Z}$ such that $um + vn = 1$. Then $u\frac{m}{n} + v = \frac{1}{n} \in A$ and hence $A \subseteq \mathbb{Z}_{S_A}$.

Ex. 15.2 $\mathbb{Z}[i]$ being an integral (commutative) domain and subring in $\mathbb{Q}[i]$ (as immediate verifications show) the ring of quotients of $\mathbb{Z}[i]$ is isomorphic with the subfield of $\mathbb{Q}[i]$ generated by $\mathbb{Z}[i]$. But this is $\mathbb{Q}[i]$ (indeed only a simple computation is needed: $\frac{a+bi}{c+di} = \frac{ac+bd}{c^2+d^2} + \frac{bc-ad}{c^2+d^2} i \in \mathbb{Q}[i]$ for $a, b, c, d \in \mathbb{Z}$ and $c + di \neq 0$).

Ex. 15.3 In 2.27 it is shown that $\mathbb{Q}^{(p)}$ is a subring in $\mathbb{Q}$ and that for each $x \in \mathbb{Q}$ one has $x \in \mathbb{Q}^{(p)}$ or $x^{-1} \in \mathbb{Q}^{(p)}$. Hence the subfield generated by $\mathbb{Q}^{(p)}$ in $\mathbb{Q}$ is $\mathbb{Q}$ itself. So the required ring of quotients is isomorphic with $\mathbb{Q}$.

Ex. 15.4 For K' the subfield generated by R in K (that is, the intersection of all the subfields of K which include R), consider $f : R \to K', f(x) = x, \forall x \in R$. By the universality property, there is a unique $\overline{f} : R_{R^*} \to K', \overline{f}\left(\frac{x}{s}\right) = x.s^{-1}$. One easily checks that $\overline{f}$ actually is a ring isomorphism (the injectivity follows directly and the surjectivity follows using the well-known form of the elements in K').

Ex. 15.5 If S is a multiplicatively closed set in a ring R and $0 \notin S$, there is a canonical bijection φ between the set of all the S-saturated ideals of R (the ideals I of R such that $I = Sat_S(I)$ where $Sat_S(I) = \{r \in R | \exists s \in S : sr \in I\}$) and all the ideals of R_S, namely $\varphi(I) = I_S = \{\frac{i}{s} | i \in I, s \in S\}$. Now, for a principal ideal $I = (a) = Ra$ one has $\varphi(I) = I_S = R_S \cdot \frac{a}{1} = \left(\frac{a}{1}\right)$, a principal ideal too.

Ex. 15.6 Only one way needs comments. If S is a saturated system clearly $R \backslash S = \bigcup \{P | P$ is a prime ideal $, P \cap S = \emptyset\}$. Conversely, if $a \in R \backslash S$ then first observe that for each unit $u \in R$, $uu^{-1} = 1$ implies $u \in S$ (a saturated system). Now, for the canonical homomorphism $f : R \to R_S$, the image $f(a) = \frac{a}{1} \in R_S$ is not a unit and so there is a prime ideal B in R_S such that $\frac{a}{1} \in B$. Hence $a \in f^{-1}(B)$ completes the proof.

Ex. 15.7 We consider the ring homomorphism $f : R[X] \to R_a$ defined by usual extension of $f(X) = a^{-1}$. Clearly, $\ker(f) \supseteq (aX - 1)$. Conversely, if a polynomial $q = \sum_n a_n X^n \in \ker(f)$ then $q(a^{-1}) = 0$ and hence $X - a^{-1}$ divides q in $R_a[X]$. So there is a polynomial $g = \sum_n g_n X^n \in R_a[X]$ such that $q = g \cdot (aX - 1)$ holds. We obtain by usual identification $-g_0 = a_0, -g_1 + ag_0 = a_1, ..$ Hence $g_n \in R$ and $q \in (aX - 1)$. It only remains to apply a ring isomorphism theorem.

Ex. 15.8 The inclusion $\mathbb{Z} \to \mathbb{Q}$ is a well-known example of epimorphism in the category of rings which is not surjective. This is a natural generalization of this example. If $\alpha, \beta : R_S \to R'$ are two ring homomorphisms and $\alpha \circ f = \beta \circ f$, for the canonical homomorphism $f : R \to R_S$ one has to prove that $\alpha = \beta$. The reader is invited to adapt 4.3.

Ex. 15.9 In other words we have to verify that any finite subset in R_S has a common denominator. Inductively write $q_1 s = r_1, .., q_{n-1} s = r_{n-1}$ and write $q_n s_n = r_n$. Then, picking $r' \in R$ and $s' \in S$ such that $s_n s' = sr'$ we obtain $r_n s' = q_n s_n s'$ and $r_i r' = q_i sr' = q_i s_n s'$ for every $1 \leq i \leq n - 1$.

Ex. 15.10 First, if f is injective, then $S \subseteq \{$ non-zero-divisors $\}$. So, f being also surjective, for each $s \in S$ surely $\frac{1}{s} \in R_S = f(R)$ so there is an element $t \in R$ such that $f(t) = \frac{t}{1} \sim \frac{1}{s}$ and hence $st = 1$ by the above inclusion. Then $S \subseteq U(R)$. Conversely, $S \subseteq U(R)$ clearly implies $S \subseteq \{$ non-zero-divisors $\}$ and so, f is injective. Finally, if $S \subseteq U(R)$ for each $\frac{a}{s} \in R_S$ we have $\frac{a}{s} = f(a.s^{-1})$ and hence f is surjective. The application to fields is obvious.

Ex. 15.11 We use the previous exercise: f is an isomorphism iff $S \subseteq U(D)$. First, if $S \subseteq U(D)$ then for each $q = \frac{a}{s} \in D_S$ we have (by natural identifications) $q = a.s^{-1}$ and $p_q = X - a.s^{-1}$. Conversely, if for each $q \in D_S$ there is a monic polynomial $p_q \in D[X]$ such that $p_q(q) = 0$ then we use $p_{\frac{1}{s}} \in D[X]$ for an arbitrary $s \in S$. If $p_{\frac{1}{s}} = X - r \in D[X]$ then $\frac{1}{s} - r = 0$ and hence $sr = 1$ or $s \in U(R)$.

Ex. 15.12 Any ideal in R_S has the form I_S where I is an ideal (see also 15.5) in R. This being finitely generated, i.e. $I = \sum_{i=1}^{n} Rx_i$, one has $I_S = \sum_{i=1}^{n} (R_S) x_i$, also a finitely generated ideal. Hence, together with R, the ring R_S is noetherian.

Ex. 15.13 It is known that $\mathbb{Z}_{\mathbb{Z}^*} \cong \mathbb{Q}$ this being the axiomatic construction of $\mathbb{Q}$ from $\mathbb{Z}$, the rational numbers being represented as fractions with integer denominators. Obviously, denominators selected from $2\mathbb{Z}^* \cup \{1\}$ suffice in order to reconstruct the whole $\mathbb{Q}$.

Ex. 15.14 If f is surjective then $R_S = f(R) = \{f(r)|r \in R\} = \{f(r.1)|r \in R\} = \{f(r).f(1)|r \in R\} = \langle f(1) \rangle$ is a cyclic R-module (via f). Conversely, if R_S is cyclic, there is an element $s \in S$ such that $R_S = \left\{f(r)\frac{1}{s}|r \in R\right\}$. Obviously, $s^2 \in S$ (being a multiplicative system) so that there is an element $t \in R$ such that $\frac{1}{s^2} \sim \frac{t}{s}$ and hence $\frac{1}{s} \sim \frac{t}{1}$. But then $f(t) = \frac{1}{s}$ and since $f(R) \supseteq f(Rt) = R_S$, the canonical homomorphism f is surjective.

 Remark. One can replace "cyclic" with "finitely generated".

Ex. 15.15 We use the previous exercise: indeed, using a well-known isomorphism theorem, it suffices to verify that f is surjective (then $f(\mathbb{Z}_n) \cong \mathbb{Z}_n/\ker(f)$). S being obviously finite, if $t = \prod_{s \in S} s$ clearly $(\mathbb{Z}_n)_S = \langle \frac{1}{t} \rangle$ is cyclic as $\mathbb{Z}_n$-module (via f). Hence f is surjective.

Ex. 15.16 The canonical homomorphism is obviously defined as follows: $f\left(\frac{(x_n)_{n\in\mathbb{N}}}{s}\right) = \left(\frac{x_n}{s}\right)_{n\in\mathbb{N}}$. Consider $\left(\frac{1}{n+1}\right)_{n\in\mathbb{N}} \in \prod_{\mathbb{N}}(\mathbb{Z})_S$. If f would be surjective, for each $n \in \mathbb{N}^*$ an element $s \in S$ and elements $y_n \in \mathbb{Z}$ should exist such that $\frac{1}{n+1} \sim \frac{y_n}{s}$ and hence $s = (n+1)\,y_n$, s would be divisible by every natural number, a contradiction.

Ex. 15.17 Clearly, $s_1 s_2 = (1+x_1)(1+x_2) = 1+x_1+x_2+x_1x_2 \in 1+I$ if $x_1, x_2 \in I$. Further, in order to verify the inclusion $I_S \subseteq Rad(R_S)$, for elements $\frac{c}{d} \in R_S$ and $\frac{a}{b} \in I_S$ we must prove that $1 + \frac{a}{b}\cdot\frac{c}{d}$ is a unit. But $1 + \frac{ac}{bd} = 1 + \frac{x}{y}$ with $x \in I$ and $y = 1+t$ with $t \in I$. Hence $1 + \frac{x}{y} = 1 + \frac{x}{1+t} = \frac{1+t+x}{1+t}$ is a unit in R_S.

Ex. 15.18 Consider the following diagram
$$\begin{array}{ccc} D & \xrightarrow{\ f\ } & Q(D) \\ \alpha\downarrow & \searrow\beta_1 & \downarrow\beta \\ D' & \xrightarrow{\ f'\ } & Q(D') \end{array}$$
where $f : D \to D'$ is an injective homomorphism of integral domains with f, f' the canonical injections. The composition $\beta_1 = f' \circ \alpha$ is an injective ring homomorphism which factors through $Q(D)$ (by the well-known universality property). The ring homomorphism β obtained in this way is clearly injective.

Ex. 15.19 Let $i : D \to D'$ be a ring isomorphism and define the map $\tilde{i} : Q(D) \to Q(D')$ by $\tilde{i}\left(\frac{a}{b}\right) = \frac{i(a)}{i(b)}$. This is well-defined because $\frac{a}{b} \sim \frac{a_1}{b_1} \Leftrightarrow ab_1 = a_1 b \Leftrightarrow i(a)i(b_1) = i(a_1)i(b) \Leftrightarrow \frac{i(a)}{i(b)} \sim \frac{i(a_1)}{i(b_1)}$. It is an injective function because one can use the above equivalence (note that i is injective) in reverse order. Moreover $\tilde{i}$ is surjective and ring homomorphism together with i.

Ex. 15.20 (a) Let P be a prime ideal in the commutative ring with identity R. Clearly, $0 \notin R\backslash P$; further, if $x, y \in R\backslash P$ then $xy \in R\backslash P$ (otherwise $xy \in P$ would imply $x \in P$ or $y \in P$).

(b) P_P is its unique maximal ideal (there is a homeomorphism (and lattice isomorphism) $\Phi : \{P \in Spec(R)|P\cap S = \emptyset\} \to Spec(S^{-1}R)$ defined by $\Phi(I) = S^{-1}I$) of R_P (obviously P is maximal relative to $P\cap(R\backslash P) = \emptyset$). Finally, R_P/P_P is isomorphic with the field of quotients of R/P. More generally, if S is a multiplicative system of $Z(R)$

(the center of R) then for each ideal I such that $I \cap S = \emptyset$ there is an isomorphism $S^{-1}R/S^{-1}I \cong (p_I(S))^{-1}(p_I(R))$ where $p_I : R \to R/I$ is the canonical projection (this is a simple application of a well-known isomorphism theorem).

Ex. 15.21 If R has no non-zero nilpotent elements, more generally, for a multiplicative system S and $\frac{a}{b} \in R_S$ let $\left(\frac{a}{b}\right)^n = 0$. There is an element $s \in S$ such that $sa^n = 0$. Hence $(sa)^n = 0$ implies $sa = 0$ and $\frac{a}{b} = 0$. Conversely, if for each prime ideal P the ring R_P has no non-zero nilpotent elements suppose that for $a \in R$ we have $a^n = 0$. Then clearly, $\left(\frac{a}{1}\right)^n = 0$ and hence for each prime ideal P there is an element $s \notin P$ such that $sa = 0$. If $Ann(a) = Ann.l(a) \cap Ann.r(a) \neq R$ then there is a maximal ideal M which includes $Ann(a)$. The ideal M being also prime there is an element $s \notin M$ with $sa = 0$, a contradiction. Hence $Ann(a) = R$ and so $a = 0$.

Ex. 15.22 This is not true. In a noetherian ring R one can show that the following conditions are equivalent:
   (i) for each maximal ideal M of R, the ring R_M is an integral domain;
   (ii) R is a finite direct product of integral domains.
   Hence it suffices to take $R = \mathbb{Z} \times \mathbb{Z}$.

Chapter 16

Rings of Continuous Functions

Ex. 16.1 (a) Straightforward verifications.

(b) and (c): one uses the constant functions.

(d) Pointwise verifications.

(e) If $R = \{0, r, ..\}$ and $X = \{a, b, ..\}$ consider $f, g : X \to R$ defined as follows:

$$f(x) = \begin{cases} 0 & \text{if } x = a \\ r & \text{if } x \neq a \end{cases}$$

and

$$g(x) = \begin{cases} r & \text{if } x = a \\ 0 & \text{if } x \neq a \end{cases}$$

These are zero divisors.

(f) Simple computations.

Ex. 16.2 The proof is straightforward.

Ex. 16.3 a) is obvious.

b) Let $\tau = \tau_0$; if $f \in C(X)$ would not be constant, then it should exist two elements $x, y \in X$ with $f(x) \neq f(y)$ and two disjoint open sets Γ and Ξ in $\mathbb{R}$ with $f(x) \in \Gamma$ and $f(y) \in \Xi$. The sets $f^{-1}(\Gamma)$, $f^{-1}(\Xi)$ should be disjoint nonvoid open sets in X, contradicting the hypothesis (τ is an indiscrete topology). Hence $C(X) = C_0(X)$.

As for the converse, case (i) see exercise 16.4 (a) or (b). For case (ii), let $X = \mathbb{Z}$ and $\tau = \{G \subset \mathbb{Z} | 0 \in G\}$. There are no disjoint open sets, therefore the only continuous functions are the constants, $C(X) = C^*(X)$.

c) Let $C(X) = C^\circ(X)$. For each $A \subset X$, the so called *characteristic function* of A, denoted $\chi_A : X \to \mathbb{R}$

$$\chi_A(x) = \begin{cases} 1 & \text{if } x \in A \\ 0 & \text{if } x \in X \backslash A \end{cases}$$

is continuous. It exists an open set Γ in $\mathbb{R}$ with $1 \in \Gamma$ and $0 \notin \Gamma$ and so $A = f^{-1}(\Gamma)$ is open. Hence the topology of X is discrete.

The converse is obvious.

Ex. 16.4 a) If $X = \{a, b\}$ is endowed with one of the following topologies $\tau_1 = \{\emptyset, \{a\}, X\}$, $\tau_2 = \{\emptyset, \{b\}, X\}$ or τ_o, then only the constant functions are continuous: $C(X) = C_o(X)$. For τ° see exercise 16.3 (c).

b) For $X = \{a, b, c\}$ the equality $C(X) = C_o(X)$ holds for most of the topologies, except for $\tau_1 = \{\emptyset, \{a\}, \{b, c\}, X\}$, and in this case the functions $f \in \mathbb{R}^X$ with $f(a) \neq f(b) = f(c)$ are also continuous. One treats similarly the topologies $\tau_2 = \{\emptyset, \{b\}, \{a, b\}, X\}$ and $\tau_3 = \{\emptyset, \{c\}, \{b, c\}, X\}$.

For the discrete topology τ° we have $C(X) = C^\circ(X)$ (see 16.3 (c)).

Ex. 16.5 It is well known that the continuity of $|f|$ follows from the continuity of f.

a) $f \geq 0$ holds iff f is a square, i.e. if it exists $g \in C(X)$ with $f = g^2$.

b) $|f|$ is the (unique) element of $C(X)$ with following two properties: (i) $|f| \geq 0$,(see (a)) and (ii) $|f|^2 = f^2$.

Ex. 16.6 $(i) \Rightarrow (ii)$ In the ring $C(X)$ the elements -1 and $\underline{1}$ are square roots of $\underline{1}$ for each space X. For G and $H = X \backslash G$, disjoint, nonvoid sets in X, consider

$$f(x) = \begin{cases} 1 & \text{if } x \in G \\ -1 & \text{if } x \in H \end{cases}$$

obviously a continuous function. f is a third square root of $\underline{1}$ in $C(X)$.

$(ii) \Rightarrow (iii)$ If f is a third square root of $\underline{1}$ ($\underline{1} \neq f \neq \underline{-1}$), then $f(x) = 1$ or $f(x) = -1$ for each $x \in X$, and $g = f \vee \underline{0}$ is an idempotent in $C(X)$; so $g, \underline{0}, \underline{1}$ are three distinct idempotent elements in $C(X)$.

$(iii) \Rightarrow (i)$ If g is an idempotent of $C(X)$ and $\underline{1} \neq g \neq \underline{0}$, then $g(x) = 1$ or 0 for each $x \in X$. The sets $\Gamma = \,]-1/2, 1/2[$ and $\Delta = \,]1/2, 3/2[$ are open and disjoint in $\mathbb{R}$. Therefore $G = g^{-1}(\Gamma)$ and $H = g^{-1}(\Delta)$ are both open, and $H = X \setminus G$.

$(i) \Rightarrow (iv)$ Let $X = X_1 \cup X_2$ and X_1, X_2 be disjoint and open sets. For $S_1 = \{f \in C(X) | x \in X_2 \Rightarrow f(x) = 0\}$ and $S_2 = \{f \in C(X) | x \in X_1 \Rightarrow f(x) = 0\}$ the equality $C(X) = S_1 \oplus S_2$ holds.

$(iv) \Rightarrow (i)$ If $C(X) = S_1 \oplus S_2$ we consider the two identities $1_{S_1} \in S_1$ and $1_{S_2} \in S_2$. From $1_{S_1} \cdot 1_{S_1} = 1_{S_1}$ follows that for $\alpha = 1_{S_1}$ the relation $\alpha^2 = \alpha$ holds. The function 1_{S_1} takes only 0 and 1 values, and therefore $X_1 = 1_{S_1}^{-1}(1) = 1_{S_1}^{-1}(]1/2, 3/2[)$ is an open subset of X. The same holds for $X_2 = 1_{S_2}^{-1}(1)$ and X_1 and X_2 are disjoint. It only remains to prove that $X = X_1 \cup X_2$. If $x \in X$ there exist two elements $f_1 \in S_1$ and $f_2 \in S_2$ whose sum is $\underline{1}$, and so either $f_1(x) \neq 0$ or $f_2(x) \neq 0$. Finally from $f_1 \cdot 1_{S_1} = f_1$ resp. $f_2 \cdot 1_{S_2} = f_2$ follows that $1_{S_1}(x) = 1$ or $1_{S_2}(x) = 1$, and so $x \in X_1 \cup X_2$.

Remark. In conditions (ii) and (iii) one can replace "three" by "four".

Ex. 16.7 Let $f \in C(X)$. The sets $G_n = \{x \in X | |f(x)| < n\}$ constitute a countable open cover of X. If $G_{n_1}, G_{n_2}, ..., G_{n_k}$ is a finite subcover, then for $n = \max\{n_1, n_2, ..., n_k\}$ and each $x \in X$ one easily checks the inequality $|f(x)| < n$. Hence f is bounded.

Ex. 16.8 We have seen in 16.3 (b), that for the space (X, τ) with $X = \mathbb{Z}$ and $\tau = \{G \subset \mathbb{Z} | 0 \in G\}$ the equality $C(X) = C_o(X)$ holds and so $C(X) = C^*(X)$ follows. This space is not countably compact: indeed, the open countable cover $\{[-n, n] | n \in \mathbb{N}\}$ has no finite subcover.

Ex. 16.9 Let
$$u(x) = \begin{cases} \frac{1}{f(x)} & \text{if } f(x) > 1 \\ 1 & \text{if } -1 \leq f(x) \leq 1 \\ -\frac{1}{f(x)} & \text{if } f(x) < -1 \end{cases}.$$
One easily checks that u has the required properties.

Ex. 16.10 Indeed, if $f \in C(X)$, one simply verifies that the equality $Z(f) = Z(|f|)$ holds.

Ex. 16.11 The following immediate equalities prove that $Z(X)$ is a sublattice (is closed under joins and meets) of the lattice of all the closed subsets of the topological space X:

$$Z(f \cdot g) = Z(f) \cup Z(g) \text{ and } Z(f^2 + g^2) = Z(f) \cap Z(g).$$

Ex. 16.12 Let us consider $X = \mathbb{Z}$ and $\tau = \{G \subset \mathbb{Z} | 0 \in G\}$. Clearly $F = \mathbb{Z}^* = \mathbb{Z} \backslash \{0\}$ is a closed subset but it is not a zero-set because $C(X) = C_0(X)$ (see 16.3 (b)).

Ex. 16.13 In a metric space (X, d) let F be a closed subset. The map $f : X \to \mathbb{R}$ defined by $f(x) = d(x, F), \forall x \in X$ is easily checked to be the required solution (i.e. $f \in C(X)$ and $Z(f) = F$).

Ex. 16.14 (a) Let $f, g \in C(X)$ be such that $f \cdot g = 1$. Then clearly $f(x) = 0$ is impossible for $x \in X$. Conversely, if $Z(f) = \emptyset$ then we can consider $g : X \to \mathbb{R}$ defined by $g(x) = \frac{1}{f(x)}$. Moreover, together with f the map $g \in C(X)$ and hence f is a unit.

(b) Let $f \in C^*(X)$. One easily verifies that f is a unit iff there is an element $\alpha \in \mathbb{R}_+^*$ (i.e. a positive real number) such that $|f| \geq \alpha$.

(c) Consider $X = \mathbb{N}^*$ with the discrete topology and $f : \mathbb{N}^* \to \mathbb{R}, f(n) = \frac{1}{n}, \forall n \in \mathbb{N}^*$. Then surely $f \in C^*(X)$ and if for a map $g \in C^*(X)$ the equality $f \cdot g = 1$ would hold, then $g(n) = n, \forall n \in \mathbb{N}^*$ and so $g \notin C^*(X)$, a contradiction.

Ex. 16.15 Consider the map $h : X \to \mathbb{R}$ defined by

$$h(x) = \begin{cases} \frac{f(x)}{g(x)} & \text{if } x \in X \backslash G \\ 0 & \text{if } x \in G \end{cases}$$

One verifies the required properties: $h \in C(X)$ and $f = g \cdot h$.

Ex. 16.16 (a) If each map from $C(X)$ has the property $\| \|$ and $f \in C^*(X)$ then there is a unit $u \in C(X)$ such that $|f| = u \cdot f$ and so for each $x \in X, f(x) > 0$ implies $u(x) = 1$ and $f(x) < 0$ implies

$u(x) = -1$. Hence $u^* = (u \wedge \underline{1}) \vee \underline{-1} \in C^*(X)$ and $|f| = u^* \cdot f$. Conversely, if each map in $C^*(X)$ has the property $\|$ and $f \in C(X)$ then for $f^* = (f \wedge \underline{1}) \vee \underline{-1}$ there is a function $u^* \in C^*(X)$ such that $|f^*| = u^* \cdot f^*$. One easily verifies $u^* \in C(X)$ and $|f| = u^* \cdot f$.

(b) All the maps have the property $\|$. Indeed, one uses the function

$$u(x) = \begin{cases} 1 & \text{if } f(x) > 0 \\ 0 & \text{if } f(x) = 0 \\ -1 & \text{if } f(x) < 0 \end{cases}$$

(c) Only the functions $f \in C(\mathbb{R})$ such that $f(x) \geq 0, \forall x \in \mathbb{R}$ or $f(x) \leq 0, \forall x \in \mathbb{R}$ satisfy the property $\|$ (one obviously uses $u = \underline{1}$ respectivelly $u = \underline{-1}$). Indeed, if f has positive and negative values, say $f(x_1) > 0$ and $f(x_2) < 0$ then f has not the property $\|$: the equalities $u(x_1) = 1$ and $u(x_2) = -1$ follow from $|f| = u \cdot f$ and so, being continuous, u must have also zero values and hence u is not a unit.

(d) In $\mathbb{R}^{\mathbb{Q}}$ consider the inclusion i (i.e. $i(x) = x, \forall x \in \mathbb{Q}$). i is clearly continuous and, if $|i| = u \cdot i$ then $u(x) = 1$ holds for $x > 0$ and $u(x) = -1$ holds for $x < 0$. Hence u is not continuous in $x = 0$.

(e) If $f \in C(\mathbb{Q})$ and $Z(f)$ is an open subset in $\mathbb{Q}$ consider $u : \mathbb{Q} \to \mathbb{R}$ defined by

$$u(x) = \begin{cases} 1 & \text{if } f(x) \geq 0 \\ -1 & \text{if } f(x) \leq 0 \end{cases}$$

First, we show that u is continuous, that is, for each open subset $\Gamma \subseteq \mathbb{R}$ the preimage $u^{-1}(\Gamma)$ is open in $\mathbb{Q}$. Indeed, we distinguish the following cases:

(i) if $\{-1, 1\} \subseteq \Gamma$ then $u^{-1}(\Gamma) = \mathbb{Q}$;

(ii) if $-1 \in \Gamma$ and $1 \notin \Gamma$ then $u^{-1}(\Gamma) = [f < 0]$ where $[f < 0] = \{x \in \mathbb{R} | f(x) < 0\}$;

(iii) if $-1 \notin \Gamma$ and $1 \in \Gamma$ then $u^{-1}(\Gamma) = [f > 0] \cup Z(f)$, and

(iv) if $1, -1 \notin \Gamma$ then $u^{-1}(\Gamma) = \emptyset$.

As for the requested example, one can take $f : \mathbb{Q} \to \mathbb{R}$,

$$f(x) = \begin{cases} 1 & \text{if } x > \sqrt{2} \\ 0 & \text{if } -\sqrt{2} < x < \sqrt{2} \\ -1 & \text{if } x < -\sqrt{2} \end{cases}$$

Then f is continuous on $\mathbf{Q}$ and $u = f$ is a unit in $C(\mathbf{Q})$ such that $|f| = u \cdot f$ holds.

Ex. 16.17 $[z_1]$ is obvious.

$[z_2]$ If $f, g \in I$, an ideal in $C(X)$, then $f^2 + g^2 \in I$ and clearly $Z(f^2 + g^2) = Z(f) \cap Z(g)$.

$[z_3]$ If $f \in I, Z = Z(f)$ and $Z \subseteq Z' = Z(g)$ then for $h = f \cdot g$ one checks $h \in I$ and $Z(h) = Z \cup Z' = Z' \in \mathcal{Z}$.

Ex. 16.18 If $Z[I] = \mathcal{Z}$ there exists a map $f \in I$ such that $Z(f) = \emptyset$. According to 16.14, (a) f is a unit and so there is a map $g \in C(X)$ such that $f \cdot g = \underline{1}$ holds. I being an ideal, $\underline{1} \in I$ and hence $\underline{1} \cdot h \in I$ for each $h \in C(X)$. So $I = C(X)$.

Ex. 16.19 Obvious.

Ex. 16.20 (a) For a maximal ideal M let $\mathcal{Z} = Z(M)$ be the corresponding proper z-filter and let $\mathcal{Z} \subseteq \mathcal{Z}_1$ be another proper z-filter. If we denote by $I = Z^{\leftarrow}(\mathcal{Z}_1)$, this is a proper ideal and $Z^{\leftarrow}(Z(M)) \subseteq I$. Clearly, $M \subseteq Z^{\leftarrow}(Z(M))$ and so $M \subseteq I$ and $M = I$. Hence $Z^{\leftarrow}(\mathcal{Z}_1) = Z^{\leftarrow}(\mathcal{Z})$ and so $\mathcal{Z}_1 = Z(Z^{\leftarrow}(\mathcal{Z}_1)) = Z(Z^{\leftarrow}(\mathcal{Z})) = \mathcal{Z}$, i.e. $\mathcal{Z}$ is a maximal z-filter.

(b) Let $\mathcal{Z}$ be a maximal z-filter and $M = Z^{\leftarrow}(\mathcal{Z})$ the corresponding proper ideal. If for another proper ideal I we have $M \subseteq I$ then $Z(I)$ is a proper z-filter and $Z(I) \supseteq Z(Z^{\leftarrow}(\mathcal{Z})) = \mathcal{Z}$. From maximality $Z(I) = Z(M)$ and so $Z^{\leftarrow}(Z(I)) = Z^{\leftarrow}(Z(M)) = Z^{\leftarrow}(Z(Z^{\leftarrow}(\mathcal{Z}))) = Z^{\leftarrow}(\mathcal{Z}) = M$. According to (a) $I \subseteq Z^{\leftarrow}(Z(I))$ and hence $I = M$, i.e. M is a maximal ideal.

Ex. 16.21 (a) Let P be a prime ideal in $C(X)$ and $f, g \in C(X)$ such that $Z(f) \cup Z(g) \in Z(P)$. We show that $Z(f) \in Z(P)$ or $Z(g) \in Z(P)$. Consider the continuous function $h = |f| - |g|$. From $(h \vee \underline{0}) \cdot (h \wedge \underline{0}) = \underline{0} \in P$ we infer $h_1 = h \vee \underline{0} \in P$ or $h_2 = h \wedge \underline{0} \in P$. If for example $h_1 \in P$ one uses in the sequel $Z_0 = Z(h_1) \in Z(P)$ and the hypothesis.

(b) Let $\mathcal{Z}$ be a prime z-filter. Then $P = Z^{\leftarrow}(\mathcal{Z})$ is a proper ideal and if $f \cdot g \in P$ we deduce $Z(f \cdot g) = Z(f) \cup Z(g) \in \mathcal{Z}$. If for instance $Z(f) \in \mathcal{Z}$ then $f \in Z^{\leftarrow}(\mathcal{Z}) = P$, and hence P is prime.

Ex. 16.22 Suppose that the z-filter $\mathcal{Z}$ is not prime. Then there are $Z_1, Z_2 \in Z[X]$ such that $Z_1 \cup Z_2 \in \mathcal{Z}$ and $Z_1 \notin \mathcal{Z}, Z_2 \notin \mathcal{Z}$ and so the set $\mathcal{Z}' = \{Z' \in Z[X] | \exists Z \in \mathcal{Z} : Z \cap Z_2 \subset Z'\}$ is a z-filter, $\mathcal{Z} \subset \mathcal{Z}'$ and $\mathcal{Z} \neq \mathcal{Z}'$ (indeed, $Z_2 \in \mathcal{Z}', Z_2 \notin \mathcal{Z}$). It is a proper one because $Z_1 \notin \mathcal{Z}'$. In fact, the hypothesis $Z_1 \in \mathcal{Z}'$ implies that $Z_1 \supset Z \cap Z_2$ for a suitable $Z \in \mathcal{Z}$. But then, from $Z_1 \cup Z_2 \in \mathcal{Z}$ and $Z \cup Z_2 \in \mathcal{Z}$ follows $Z_1 = (Z \cap Z_2) \cup Z_1 = (Z \cup Z_1) \cap (Z_2 \cup Z_1) \in \mathcal{Z}$ and so $\mathcal{Z}$ is not maximal.

Ex. 16.23 (a) Notice that $g \in I_2$ means that $g(x) = x^2 f(x), \forall x \in \mathbb{R}$ for some $f \in C(X)$. But then $g(x) = x f_1(x)$ where $f_1(x) = x f(x)$ and $f_1 \in C(X)$. Hence $g \in I_1$. Clearly $1_{\mathbb{R}} \in I_1$ but $1_{\mathbb{R}} \notin I_2$ because its derivative takes the value 1 for $x = 0$ and the derivative of each element from I_2 has the value zero in $x = 0$ (indeed if $f(x) = x^2 h(x)$ with $h \in C(X)$ then $\lim_{x \to 0} \dfrac{f(x) - f(0)}{x} = \lim_{x \to 0} x h(x) = \lim_{x \to 0} x \cdot \lim_{x \to 0} h(x) = 0.h(0) = 0$). Hence $I_1 \neq I_2$.

Now, $Z(I_1) = Z(I_2)$ because both contain only the closed subsets $\Gamma \subset \mathbb{R}$ which include 0. Indeed, for such a subset Γ let $h(x)$ be the distance between x and Γ. Then for $f(x) = x h(x)$ and $g(x) = x^2 h(x)$ one checks $f \in I_1, g \in I_2$ and $Z(f) = Z(g) = \Gamma$.

(b) It suffices to observe that $1_{\mathbb{R}} \cdot 1_{\mathbb{R}} \in I_2$ but $1_{\mathbb{R}} \notin I_2$ respectively $\left(1_{\mathbb{R}}\right)^{\frac{1}{3}} \cdot \left(1_{\mathbb{R}}\right)^{\frac{2}{3}} \in I_1$ but none of these factors does belong to I_1.

(c) $M_0 = \{f \in C(X) | f(0) = 0\}$ obviously includes I_1 and I_2. In order to prove that it is also a maximal ideal, let I be an ideal in $C(X)$ such that $M_0 \subset I$ and $M_0 \neq I$. Using a map $f \in I \backslash M_0$ from $f(0) \neq 0$ we infer $Z\left(f^2 + 1_{\mathbb{R}}^2\right) = Z(f) \cap Z(1_{\mathbb{R}}) = \emptyset$ and so $f^2 + 1_{\mathbb{R}}^2 \in I$. Using the exercise 16.14 (a) $f^2 + 1_{\mathbb{R}}^2$ is a unit and so $I = C(X)$.

(d) We already saw in (a) that $Z(I_1) = \{\Gamma \subset \mathbb{R} | \Gamma \in \mathcal{F}, \underline{0} \in \Gamma\}$. If $\mathcal{Z}$ is a z-filter such that $Z(I_1) \subset \mathcal{Z}$ and $Z(I_1) \neq \mathcal{Z}$, choose $Z \in \mathcal{Z} \backslash Z(I_1)$. From $0 \notin Z$ and $\{0\} \in Z(I_1)$ we obtain $\{0\} \cap Z = \emptyset$ included in $\mathcal{Z}$ and so $\mathcal{Z}$ is not a proper z-filter. Hence $Z(I_1)$ is maximal.

Remark. If $Z(I)$ is a maximal z-filter, I must not be a maximal ideal.

(e) $Z^{\leftarrow}(Z(I_1))$ is the maximal ideal from (c).

Ex: 16.24 (a) The first part is straightforward. $\mathcal{Z}_+$ is not prime because it includes not $Z_1 = [0,1]\setminus \bigcup_{n\in\mathbb{N}^*}]\frac{1}{2n}, \frac{1}{2n-1}[$ and

$Z_2 = [0,1]\setminus \bigcup_{n\in\mathbb{N}^*}]\frac{1}{2n+1}, \frac{1}{2n}[$ but it includes $Z_1 \cup Z_2$. One similarly treats $\mathcal{Z}_-$.

(b) $\mathcal{Z}_0$ is equal to $Z(I_1)$ from the previous exercise. It is maximal and hence prime.

Chapter 17

Special problems

Ex. 17.1 (a) One checks that $(a^{-1} - b^{-1})^{-1} = -a(a-b)^{-1}b$ holds.
  (b) Indeed, $-a(a-b)^{-1}b(a^{-1} - b^{-1}) =$
  $-a(a-b)^{-1}ba^{-1} + a(a-b)^{-1} = a(a-b)^{-1}(1 - ba^{-1}) =$
  $a(a-b)^{-1}(aa^{-1} - ba^{-1}) = a(a-b)^{-1}(a-b)a^{-1} = 1$. Similarly
$-(a^{-1} - b^{-1})a(a-b)^{-1}b = 1$.

Ex. 17.2 First we verify that for every $r, s \in R$ the equality $r^2sr^2 = r^2s$ holds. Indeed, simple computations show that $(r^2sr^2 - r^2s)^2 = 0$ and so $r^2sr^2 - r^2s = (r^2sr^2 - r^2s)^3 = 0$. Similarly $r^2sr^2 = sr^2$ and hence $r^2s = sr^2$ or $\{r^2 | r \in R\} \subseteq Z(R)$. Finally, $rs = r^3s^3 = r(r^2s^3) = rs^3r^2 = rs^2sr^2 = s^2rsr^2 = s(sr)(sr)r$
  $= s(sr)^2r = (sr)^2sr = (sr)^3 = sr$ for every $r, s \in R$.

Ex. 17.3 If the ring has identity this was 1.31. We obtain $char(R) = 2$ identically. Now, $x + x^2 = (x + x^2)^6 = x^6 + 6x^7 + 15x^8 + 20x^9 + 15x^{10} + 6x^{11} + x^{12} = x + x^3 + x^5 + x^2$ so that $x^3 + x^5 = 0$ or $x^5 = -x^3 = x^3$. Multiplying by x we have also $x = x^6 = x^4$ and hence $x^2 = x^5$ or $x^2 = x^3$. Finally, $x = x^6 = x^2.x^4 = x^2.x = x^3 = x^2$.

Ex. 17.4 (a) Indeed, $(a_i..a_na_1..a_{i-1})^2 = a_i..a_n(a_1..a_n)a_1..a_{i-1} = a_i..a_n.0.a_1..a_{i-1} = 0$.
  (b) We use an induction on k. Indeed, if $k = 1$ then
  $(a_1b_1a_2a_3..a_n)^2 = a_1b_1(a_2a_3..a_na_1)b_1a_2..a_n = a_1b_1.0.b_1..a_n = 0$ (using (a)). Further, if the statement is true for $k - 1$ we compute

$\left(a_1b_1..a_kb_ka_{k+1}..a_n\right)^2 =$
$a_1b_1..a_kb_k\left(a_{k+1}..a_na_1b_1a_2..b_{k-1}a_k\right)b_ka_{k+1}..a_n = 0$ the bracketts containing a zero product (one has to apply (a) to the induction hypothesis). So the statement holds also for k.

(c) Each permutation being a product of transpositions it suffices to prove the assertion for a transposition $\sigma = (ij)$, i.e.

$P = a_1..a_{i-1}a_ja_{i+1}..a_{j-1}a_ia_{j+1}..a_n = 0$. What follows is a little bit harder: we compute the $j-1$ power of this last product, choose $j-1$ suitable elements b and apply (b). Indeed, each element b (excepting b_{i-1}, b_i and b_{j-1}, b_j) has n factors (the exceptions have only $j-i$ resp. $n-j+i$ factors) from P (in fact a "circular" permutation of it): if we choose $b_1 = a_2..a_na_1, b_2 = a_3..a_1a_2, .., b_{i-1} = a_ja_{i+1}..a_{j-1}, b_i = a_{j+1}..a_{i-1}a_j$ and $b_{i+1} = a_{i+2}..a_ja_{i+1}, .., b_{j-1} =$
$a_ia_{j+1}..a_{i-1}, b_j = a_{i+1}..a_{j-1}a_i$ then $(P)^{j-1} = a_1b_1..a_jb_ja_{j+1}..a_n = 0$
and hence $P = 0$.

Remark. If R has identity the cube P^3 suffices for the proof, with only 4 suitable elements b.

Ex. 17.5 Let $x, y \in R$. From $x^{n+1}y^{n+1} = (xy)^{n+1} = (xy)^nxy = x^ny^nxy$ we obtain

$$x^n(xy^n - y^nx)y = 0. \tag{17.1}$$

Replacing in (1) x by $x+1$ we obtain

$$(x+1)^n(xy^n - y^nx)y = 0. \tag{17.2}$$

Developing $(x+1)^n$ binomially (the commutativity is not needed), multiplying (2) to the left with x^{n-1} and using again (1) we deduce

$$x^{n-1}(xy^n - y^nx)y = 0. \tag{17.3}$$

Repeating the (1) $\Rightarrow$ (3) procedure we now get step by step

$$x(xy^n - y^nx)y = 0 \tag{17.4}$$

and even

$$(xy^n - y^nx)y = 0. \tag{17.5}$$

We now begin with $x^{n+2}y^{n+2} = (xy)^{n+2} = (xy)^{n+1}xy = x^{n+1}y^{n+1}xy$ and obtain

$$x^{n+1}(xy^{n+1} - y^{n+1}x)y = 0. \qquad (17.6)$$

Repeating the technique $(1) \Rightarrow (5)$ we now obtain

$$(xy^{n+1} - y^{n+1}\overset{..}{x})y = 0. \qquad (17.7)$$

But (5) implies $xy^{n+1} = y^n xy$ or

$$yxy^{n+1} = y^{n+1}xy. \qquad (17.8)$$

From (7) and (8) follows $(xy - yx)y^{n+1} = 0$; the replacement of y by $y + 1$ and a repeated application of the $(1) \Rightarrow (5)$ technique leads to $(xy - yx)y = 0$ and finally to $xy - yx = 0$. Hence R is commutative.

Ex. 17.6 The solution needs knowledge of field extensions (which are not treated in this collection). Briefly, the proof relies on a lemma due to Kaplansky: if $K \neq F$ is an extension field of F and for each given $a \in K$ there is a $n(a) \in \mathbb{N}^*$ such that $a^{n(a)} \in F$ then either K is purely inseparable over F or, K is of prime characteristics and is algebraic over its prime field P, and a result concerning algebras established by Noether and extended by Jacobson: if D is a noncommutative division algebra algebraic over its center Z then there is an element in D, not in Z, separable over Z. The proof is divided into three cases: for division rings, for primitive rings and for the general case where R is represented as a subdirect sum of prime rings so that the proof finally reduces to prime rings. For a complete treatment see [7], p.79, Theorem 3.2.2.

Ex. 17.7 More generally, one can prove the following statement: if a ring R has a left (or right) nonzero divisor $a \in R$ then $char(R) = ord_{(R,+)}(a)$. Indeed, if $ord_{(R,+)} = \infty$ nothing is to be proved. If $ord_{(R,+)}(a) = n$ then $\forall r \in R : 0 = (na)r = a(nr) \Rightarrow nr = 0$, a being no left zero divisor. We deduce from this result that all nonzero elements in a field K have the same order in the additive group $(K, +)$. Hence $2a = 3b = 0$ for nonzero elements in K is not possible.

Ex. 17.8 First observe that for each idempotent element $e \in R, 1 - e$ is also idempotent. $1 \neq 0$ implies that $e \neq 1 - e$. So we can group the nonzero elements in pairs, but $e(1 - e) = 0$ and hence:

(i) if 1 is the only nonzero idempotent element of R, then the required product is evidently 1;

(ii) if there is an idempotent $e \notin \{0, 1\}$ then the product is 0.

Ex. 17.9 If $I^2 \neq 0$ there is an element $a \in I, a \neq 0$ such that the right ideal $aI = \{ax | x \in I\} \neq 0$. I being a minimal right ideal we must have $aI = I$ and hence there is an element $e \in I$ such that $ae = a$. One easily gets $a(e^2 - e) = 0$ and so $e^2 - e \in Ann.r(a) \cap I$ where $Ann.r(a) = \{r \in R | ar = 0\}$ is the right annihilator of a in R. But $Ann.r(a) \cap I \neq I$ because $aI \neq 0$ and hence $Ann.r(a) \cap I = 0$ and $e^2 - e = 0$, e is idempotent. From $a \neq 0$ we get $e \neq 0$ and, I being minimal, we obtain $I = eR$.

Ex. 17.10 First observe that if e is an idempotent element of R then eRe is a subring of R with the identity $e \in eRe$ (no identity in R is needed: $e = eee \in eRe$). Next, if $ere \in eRe$ is an idempotent then $e - ere$ is also an idempotent (by straightforward computation), obviously orthogonal with ere. Now if e is primitive then from $e = ere + (e - ere)$ follows that $ere \in \{0, e\}$. Conversely, if e is not primitive, let $e = u + v, u \neq 0 \neq v, u^2 = u, v^2 = v, uv = 0, vu = 0$. Then eue is idempotent in eRe and $eue \neq 0$ (indeed from $eue = 0$ we infer $e = eee = e(u + v)e = eue + eve = eve$ and then $u = u^2 = (e - v)^2 = e^2 - ve - ev + v^2 = 0$ because $ev = (u + v)v = v$ and similarly $ve = v$ or $e = eve = ev$).

Ex. 17.11 As in the previous exercise, eRe is a subring of R with identity e. If $exe \neq 0$ then, R being regular, let $y \in R$ such that $exe = (exe)y(exe)$. Then $(exe)(eye) = exeye$ is idempotent : indeed

$(exe)(eye)(exe)(eye) = (exeyexe)ye = exeye$. Using the previous exercise (e is primitive) $(exe)(eye) \in \{0, e\}$ and hence $(exe)(eye) = e$ (indeed $(exe)(eye) = 0$ implies $exe = (exe)y(exe) = 0$), the identity in eRe. So exe is a left unit and the rest follows symmetrically.

Ex. 17.12 First observe that there is a left ideal $M \leq N$ such that $0 \neq M^2 = M$. Indeed, R being left artinian the descending chain of left ideals $N \supseteq N^2 \supseteq .. \supseteq N^k = N^{k+1} = ..$ becomes stationary and $M = N^k$. Further, choose a left ideal $L \leq M$ minimal with respect to the property $M \cdot L \neq 0$. This can be done using the descending chain condition on the set of all the left ideals $\mathcal{A} = \{A | M \cdot A \neq 0\}$.

Next, we show that there is an element $b \in L$ such that $Mb = L$. Indeed, since $M \cdot L \neq 0$ there is an element $b \in L$ such that $Mb \neq 0$ and then $0 \neq Mb = M^2b = M(Mb)$. Hence $Mb \in \mathcal{A}$ and by minimality $Mb = L$.

Finally, there is an element $a \in M \leq N$ such that $ab = b$ (from $Mb = L$). This is the required element. Indeed, if a is nilpotent, that is, $a^s = 0, s \in \mathbb{N}^*$, then $b = ab = a^2b = .. = a^sb = 0$, which contradicts the selection of L.

Ex. 17.13 Consider $P \in K[X]$ defined as follows

$$P(X) = \sum_{a \in K} \left(f(a) \prod_{b \in K \setminus \{a\}} \frac{X - b}{a - b} \right). \text{ Obviously, } f(a) = P(a), \forall a \in$$

K.

Ex. 17.14 (i)$\Rightarrow$(ii) If $a \in R$ and $b \in R : aba = a$ then $e = ba$ is an idempotent and then $Re \subseteq Ra$. Moreover, since $a = ae \in Re$ we have $Re = Ra$ and hence $R = Re \oplus R(1 - e) = Ra \oplus R(1 - e)$.

(ii)$\Rightarrow$(i) If $a \in R$, by hypothesis there is a left ideal I such that $R = Ra \oplus I$ and hence $1 = ba + x$ with $b \in R, x \in I$. But then $a = aba + ax = aba$ because $0 = ax \in Ra \cap I$.

(ii)$\Rightarrow$(iii) By hypothesis it suffices to show that for every two idempotents $e, e' \in R$ the sum $Re + Re'$ is generated also by a single idempotent. First observe that $Re + Re' = Re + R(e' - e'e)$. Then choose an element $b \in R$ such that $e' - e'e = (e' - e'e)b(e' - e'e)$ ((i) and (ii) are already verified to be equivalent). If u denotes $b(e' - e'e)$ one checks $u^2 = u$ and $ue = 0$. Hence $Re + Re' = Re + Ru = R(e + u - eu)$ with an idempotent $e + u - eu$.

(iii)$\Rightarrow$(ii) is obvious.

Ex. 17.15 (a) If for instance $ac = 0$ or $ca = 0$ one deduces $a = a(b+c)a$ and so by uniqueness $b = b + c$ holds. Hence $c = 0$.

(b) From $aba = a$ we deduce $abab = ab$ or $a(bab - b) = 0$. Using (a) we get $bab = b$.

(c) For any $a \in R, a \neq 0$ denote $e = ab$. This is the identity of R: for every $c \in R$ we have $(ce - c)e = cabab - cab = cab - cab = 0$. Again by (a) $e \neq 0$ and hence $ce = c$. Similarly, $ec = c$.

Remark. In an analogous manner one proves ba is the identity and so $ba = e$ (by the uniqueness of identity).

(d) For each $c \neq 0$ if $c = cdc$ then (as we have seen in (c)) $e = cd$ is the identity. Using the above remark, $e = dc$ and so c has inverse.

Ex. 17.16 If the ring R is regular, for each $a \in R$ there is an element $b \in R$ such that $a = aba$. We take $c = bab$ and then $cac = b(aba)bab = babab = bab = c$. As for the application, let $a \in Z(R)$ for a regular ring R. As we have seen there is an element $c \in R$ such that $a = aca, c = cac$ and we check that also $c \in Z(R)$. For each $r \in R$ using $a \in Z(R)$ and $a = aba = a^2b$ the following hold: (a) $babr = (babr)ba = (ba)^2rb = barb$, (b) $r(ba) - (ba)r = (rba - bar)ba = rba - ba^2rb = (rb - br)a = 0$. Finally,
$$rc - cr = r(bab) - (bab)r \overset{(a)}{=} rbab - barb = (rba - bar)b \overset{(b)}{=} 0.$$

Ex. 17.17 Let $R = \sum_{i \in I} A_i$ with A_i ideals in R. Then $1 \in \sum_{i \in I} A_i$ and hence there are $\{i_1, .., i_k\} \subseteq I$ such that $1 = a_{i_1} + .. + a_{i_k}$ with suitable $a_{i_j} \in A_{i_j}$. But then $R = \sum_{j=1}^{k} A_{i_j}$.

Ex. 17.18 We know that $\mathbb{Z}[i]$ is an Euclidean ring (see 6.5) together with the norm N. Let $a + bi$ be a nonzero element from $\mathbb{Z}[i]$ and $H = (a + bi)$. For each element $z \in \mathbb{Z}[i]$ there are elements q and r such that $z = (a + bi)q + r$ where $r = 0$ or $N(r) < a^2 + b^2$. Hence, each nonzero element in $\mathbb{Z}[i]/H$ has the form $r + H$ where $N(r) < a^2 + b^2$. But the set $M = \{x \in \mathbb{Z}[i] | N(x) < a^2 + b^2\}$ is obviously finite, and then so is also $\mathbb{Z}[i]/H$.

Ex. 17.19 $(b) \Rightarrow (c)$ is obvious: $(0,0) \neq (1,0) \neq (1,1)$ in $U \times V$ and $(1,0)^2 = (1,0)$ is the required idempotent in $U \times V \cong R$.

$(c) \Rightarrow (b)$ If $e^2 = e$ is idempotent in R and $e \notin \{0,1\}$ then Re and $R(1 - e)$ are proper ideals in R and even comaximal (see 2.24) that is $Re + R(1-e) = R$. But $Re \cap R(1-e) = (0)$ and so $R \cong R/(Re \cap R(1 - e)) \cong R/Re \times R/R(1 - e)$ the required proper decomposition.

$(a) \Rightarrow (c)$ We consider again (see also 13.11 and 13.12) $\mathcal{R}(R)$, the nilradical of R, that is $\cap\{P|P \in Spec(R)\}$, the smallest ideal (in a commutative ring) such that $\mathcal{R}(R/\mathcal{R}(R)) = (0)$. We split our proof into two cases. First, suppose that $\mathcal{R}(R) = (0)$. $Spec(R)$ being not connected, there are F_1, F_2 closed, disjoint, proper subsets of $Spec(R)$ such that $Spec(R) = F_1 \cup F_2$. Let $I_i = \cap\{P|P \in F_i\}, i \in \{1,2\}$; F_i being closed subsets we have $F_i = \overline{F_i} = V(I_i)$. It is easily checked that there is no $P \in Spec(R)$ such that $I_1 + I_2 \subset P$ and hence $I_1 + I_2 = R$. So $\mathcal{R}(R) = (0)$ implies $I_1 \cap I_2 = (0)$ with $I_1 \neq (0) \neq I_2$ and hence $R \cong R/I_1 \times R/I_2$. Next, suppose that $\mathcal{R}(R) \neq (0)$ and consider $R/\mathcal{R}(R)$ ring with zero nil (or prime) radical. Each prime ideal includes $\mathcal{R}(R)$ so that there is a topological bijection between $Spec(R)$ and $Spec(R/\mathcal{R}(R))$ and hence $Spec(R/\mathcal{R}(R))$ is not connected. Using the first case, there are rings $U \neq 0 \neq V$ such that $R/\mathcal{R}(R) \cong U \times V$. Using $(b) \Rightarrow (c)$ let $\overline{x} \notin \{\overline{0}, \overline{1}\}$ an idempotent element in $R/\mathcal{R}(R)$. "Lifting" this idempotent (see 2.19) in R, $(c) \Rightarrow (b)$ ends this proof.

$(c) \Rightarrow (a)$ We have $V(e) \cup V(1-e) = V(e.(1-e)) = V(0) = Spec(R)$ and hence, $0 \neq e \neq 1$ implies that $V(e)$ and $V(1-e)$ are closed subsets that differ from $\emptyset$ and $Spec(R)$.

Ex. 17.20 The map u is evidently injective. It is also surjective iff (b) takes place (there is a solution r for the system of congruences modulo the ideals I_i). The rest is a generalisation of 2.24.

Ex. 17.21 Together with R, $R[X]$ is a unique factorization domain and so there is an irreducible element $g \in R[X], g = b_0 + b_1 X + .. + b_m X^m$ which divides f and p divides b_0. Let us denote the polynomial $h = c_0 + c_1 X + .. + c_l X^l$ such that $f = g \cdot h$. Now p divides $a_0 = b_0 c_0$ and p^2 divides not a_0 so that p divides not c_0. Let $k \in \mathbb{N}$ be minimal with the property that p divides not b_k. But p divides not $a_n = b_m c_l$ and hence $k \leq m \leq n$. Surely $a_k = b_k c_0 + b_{k-1} c_1 + .. + b_0 c_k$ and so (by the minimality of k) p divides $a_k - b_k c_0$. But p divides not $b_k c_0$ and by the hypothesis $k = n, l = 0$. Hence in $R[X]$ we have $f = c_0 g$ with $c_0 \in R$ and $g \in R[X]$ irreducible.

In $K[X]$ suppose that we have the decomposition $f = \tilde{f}_1 .. \tilde{f}_m$. For a suitable element $t \in R$, all the polynomials $t\tilde{f}_s \in R[X]$ and $c_0 t^m g = \left(t\tilde{f}_1\right) .. \left(t\tilde{f}_m\right)$. $R[X]$ being a unique factorization domain g divides one of the terms, say $t\tilde{f}_j$. Hence there is an element $u \in R$ such that $ug = t\hat{f}_j$ and so $f = \left(c_0 \frac{t}{u}\right)\tilde{f}_j$ is irreducible in $K[X]$.

As for the final remark one uses the following result: for a unique factorization domain R and K his field of quotients, a polynomial f is irreducible in $R[X]$ iff f is primitive and irreducible in $K[X]$.

Bibliography

[1] Anderson F.W.,Fuller K.R., *Rings and Categories of Modules*, Graduate Text in Mathematics, Springer, 1974.

[2] Călugăreanu G., *Collection of Problems in Algebra*, Cluj-Napoca University Press, 1978 (Romanian).

[3] Faith C., *Algebra: Rings, Modules and Categories I*, Springer Verlag, 1973; *Algebra II: Ring Theory*, Springer Verlag, 1976.

[4] Franciosi S., de Giovanni F., *Esercizi di algebra*, Aracne, 1993 (Italian).

[5] Gillman L., Jerison M., *Rings of Continuous Functions*, Van Nostrand, Princeton, 1960.

[6] Gray M., *A radical Approach to Algebra*, Addison-Wesley, 1970.

[7] Herstein I.N., *Noncommutative Rings*, Wiley and Sons, 1968.

[8] Herstein I.N., *Topics in Ring Theory*, Univ. of Chicago Press, 1969.

[9] Hocquemiller J., Weil J., *Algèbre - S.MacLane, G.Birkhoff: Solutions dévelopées des exercices, 1re partie*, Gauthier-Villars, 1972 (French).

[10] Ion D.I., Radu N., Niță C., Popescu D., *Problems in Algebra*, Didactic and Pedagogic Press, Bucharest, 1981 (Romanian).

[11] Jans J.P., *Rings and Homology*, Holt, Rinehart, Winston, 1964.

[12] Kaplansky I., *Commutative Rings*, Allyn and Bacon, 1970.

[13] Kaplansky I., *Fields and Rings*, Univ. of Chicago Press, 1969.

[14] Kasch F., *Moduln und Ringe*, B.G.Teubner Stuttgart, 1977 (German).

[15] Kelley J.L., *General Topology*, Van Nostrand, Princeton, 1955.

[16] Kertesz A., *Lectures on Artinian Rings*, Akademiai Kiado, Budapest, 1987.

[17] Lambek J., *Lectures on Rings and Modules*, Blaisdell, 1966.

[18] McCoy H.N., *The Theory of Rings*, MacMillan Co., 1964.

[19] Meyberg K., Vachenauer P., *Aufgaben und Lösungen zur Algebra*, Hanser Verlag, München, 1978 (German).

[20] Năstăsescu C., *Rings.Modules.Categories*, Academic Press (of Romania), 1976 (Romanian).

[21] Năstăsescu C., Andrei G., Ţena M., Otărăşanu, *Problems of Algebraic Structures*, Academic Press (of Romania), 1988 (Romanian).

[22] Niţă C., Spîrcu T., *Problems of Algebraic Structures*, Technical Press, Bucharest, 1974 (Romanian).

[23] Pumplün D., Sidow W., *Lineare Algebra I.* Fernuniversität-GH Hagen, 1989 (German).

[24] Rădescu E., Rădescu N., Vraciu G., *Algebra, Collection of Problems*, Craiova University Press, 1975 (Romanian).

[25] Rowen L.H., *Ring Theory*, vol.1+2, Academic Press, 1988.

[26] Spîrcu T., *Algebraic Structures through Problems*, Scientific Press, Bucharest, 1991. (Romanian).

[27] Stenström B., *Rings of Quotients*, Springer Verlag, 1975.

[28] Wisbauer R., *Foundations of Module and Ring Theory*, Gordon and Breach Science Publishers, 1991.

Index

Kluwer Texts in the Mathematical Sciences

KLUWER ACADEMIC PUBLISHERS – DORDRECHT / BOSTON / LONDON